A guidebook to
microscopical methods

A guidebook to microscopical methods

A. V. Grimstone
*Fellow of Pembroke College and Assistant Director of Research
in Zoology in the University of Cambridge*

and R. J. Skaer
*Fellow of Peterhouse and Assistant Director of Research in Medicine
in the University of Cambridge*

Cambridge at the University Press 1972

Published by the Syndics of the Cambridge University Press
Bentley House, 200 Euston Road, London NW1 2DB
American Branch: 32 East 57th Street, New York, N.Y.10022

© Cambridge University Press 1972

Library of Congress Catalogue Card Number: 70–182027

ISBNS:
0 521 08445 8 hard covers
0 521 09700 2 paperback

Printed in Great Britain
at the University Printing House, Cambridge
(Brooke Crutchley, University Printer)

Contents

'We'll try it,' the professor said to me, grimly, 'with every adjustment of the microscope known to man. As God is my witness, I'll arrange this glass so that you see cells through it or I'll give up teaching. In twenty-two years of botany, I – ' He cut off abruptly for he was beginning to quiver all over, like Lionel Barrymore, and he genuinely wished to hold onto his temper; his scenes with me had taken a great deal out of him.

So we tried it with every adjustment of the microscope known to man. With only one of them did I see anything but blackness or the familiar lacteal opacity, and that time I saw, to my pleasure and amazement, a variegated constellation of flecks, specks and dots. These I hastily drew. The instructor, noting my activity, came back from an adjoining desk, a smile on his lips and his eyebrows high in hope. He looked at my cell drawing. 'What's that?' he demanded, with a hint of a squeal in his voice. 'That's what I saw,' I said. 'You didn't, you didn't, you *did*n't!' he screamed, losing control of his temper instantly, and he bent over and squinted into the microscope. His head snapped up. 'That's your eye!' he shouted. 'You've fixed the lens so that it reflects! You've drawn your eye!'

– James Thurber, *My Life and Hard Times*

Preface

Like its predecessor, Pantin's *Notes on Microscopical Technique for Zoologists*, this book is intended primarily for students and those beginning research. We hope it may also be useful to more experienced workers needing information about techniques with which they are unfamiliar. As the title implies, our aim has been to provide a guidebook, enabling the reader to find his way through the large and often bewildering array of methods now described in the literature. We have tried firstly to single out the more important and commonly used techniques and to describe them with sufficient practical details to enable them to be carried out without further instructions, and secondly to give notes and references to permit the reader to obtain additional information, both about these techniques and about ones less frequently used.

Compared with Pantin's book, the range of techniques covered here is somewhat wider. In particular, there are more methods for the study of cells, and we have not confined ourselves to animal material. These differences, of course, reflect the present trends of biological research.

We offer our thanks to several colleagues who have been kind enough to give us the benefit of their advice and criticism on various sections. In particular, we would acknowledge gratefully the help of Dr W. B. Amos, Dr M. Ashburner, Dr B. L. Gupta, Dr A. M. Mullinger, Professor J. A. Ramsay and Dr J. Smart.

We should be grateful if users of this book would bring to our notice any errors they may detect, in case a subsequent edition should be called for.

Preliminaries

(1) As a general rule, use simple techniques in preference to complicated ones. Whenever possible, use living material.

(2) If a method does not give satisfactory results at the first attempt it is often worth trying it again, before abandoning it. Experience and practice are important in microscopy, and results often improve with successive attempts even though one may not be aware of doing anything different.

(3) Attention to detail is important. Solutions should be made up accurately, and used fresh if unstable. Glassware must be clean. Handle slides with forceps and keep fingers out of solutions.

(4) In the recipes given here (and in other books) precise times are usually given for the various steps in a procedure. Often these can be varied over a considerable range without much affecting the results. Only experience will show what variation is possible, however, and until this is gained it is recommended that the standard times given here should be adhered to.

(5) The toxicity of many of the reagents used in microscopy is commonly not appreciated. *In general, avoid imbibing, inhaling or absorbing any kind of chemical or solution.* In addition to such obviously toxic reagents as mercuric chloride or osmium tetroxide, such substances as basic fuchsin, benzene, cresol, glutaraldehyde, methacrylates, propylene oxide or epoxy-resins should all be treated with the greatest respect.

Methods of observation

For the principles and methods of use of all kinds of microscope see Ruthmann (1970) or Slayter (1970). For electron microscopy see p. 59.

Bright-field microscope

For a simple account of the use of the ordinary light microscope see Barer (1968); for advanced treatments of image formation, diffraction theory and resolution see Martin (1966) or Michel (1964).

The *resolution* (δ) is defined as the smallest distance between two details in the specimen which can be distinguished as separate. It is given approximately by the formula

$$\delta = \frac{0.6\lambda}{n\sin\alpha}$$

where λ is the wavelength of the light used to illumine the specimen, n is the refractive index of the medium between the objective and the specimen and α is half the angle subtended by the object at the surface of the objective. The quantity $n \sin \alpha$ is called the *numerical aperture* of the objective. The best resolution obtainable with the light microscope is approximately 0.2 μm.

The different kinds of objective differ principally in the extent to which they are corrected for chromatic aberration and curvature of field. Achromatic objectives are reasonably well corrected for chromatic aberration over the range of wavelengths to which the eye is most sensitive. They are not well corrected for the shorter blue or longer red wavelengths and hence not generally suitable for photomicrography. Fluorite objectives are better corrected and image contrast is good, making them suitable for photomicrography. Apochromatic objectives are more complex in construction, more expensive, and still better corrected. There may be some loss of image contrast in these. Planachromat and planapochromat objectives are additionally corrected for curvature of field.

The magnification and numerical aperture of an objective are usually engraved on it (e.g. 40/0.75).

Objectives are usually designed for use with coverslips 0.17 mm thick (No. 1). Commercially available coverslips vary considerably in thickness and this may cause appreciable deterioration in image quality, especially with high-power, dry objectives. Some such objectives have a correction collar to compensate for variation in coverslip thickness. Note that a thick layer of mounting medium may act like a thick coverslip.

The function of the *condenser* is to focus an image of the light source on the specimen. The condenser is always used in conjunction with a plane mirror: the concave mirror provided on some microscopes is for use only in very low-power work without a condenser. Correct focusing of the condenser is essential. In most microscopes an iris diaphragm (the *field iris*) is mounted in front of the lamp. The condenser is focused so as to form a sharp image of the edge of this diaphragm on the specimen. The diaphragm is then opened just sufficiently to fill the field of view. If it is opened further glare will result. The *condenser iris* is then opened just sufficiently far to fill about two-thirds of the back focal plane of the objective with light. This can best be done by removing the eyepiece and inspecting the back lens of the objective directly while adjusting the condenser iris. The *intensity of illumination* is adjusted either by controlling the current through the lamp (usually by means of a variable transformer) or by using neutral filters. It must *not* be controlled by adjusting the condenser or field irises or by defocusing the condenser: glare and loss of resolution will result. The numerical aperture of the condenser may limit resolution and it is essential to oil the condenser to the slide for work at the highest resolution. Like objectives, condensers are corrected to different extents for curvature of field and chromatic aberration.

Oculars (eyepieces) are of two main types, Huygens and compensating. The former are cheaper and suitable for use only with achromatic objectives. Oculars may be designed specifically for use with the objectives produced by the same manufacturer. The magnification of oculars should be such as to result in a total magnification 500–1000 times that of the objective numerical aperture.

For cleaning lenses use a dry camel-hair brush or a jet of dry air for removing dust, and use lens tissue moistened with the least possible quantity of xylene for removing immersion oil or grease. Clean cautiously. Do not rub the surface of a dirty lens when it is dry, otherwise it may be scratched. For final polishing breathe on the lens and rub gently with lens tissue.

Dark-field microscope

In this the specimen is illuminated by a hollow cone of light, usually obtained with a special condenser. Objects which scatter light appear bright against a dark background. Such scattering depends on the existence of a sharp change of refractive index at the surface of the object. Very small particles can be made visible by this means, and the method is useful for studying flagella, sperm tails, etc. For high-power dark-field work the condenser must be oiled to the slide. The effective numerical aperture of the objective must be low enough to exclude direct light from the condenser; for oil-immersion work this is achieved by using an objective with a built-in iris diaphragm, by means of which the numerical aperture can be reduced to about 1.0. For low magnifications dark-field illumination can be improvised on an ordinary phase-contrast microscope by using the condenser phase-plate intended for the × 100 objective in conjunction with a low-power objective.

Phase-contrast microscope

For the theory see Barer (1966) or Ross (1967). The phase-contrast microscope is used chiefly for examination of living cells. It makes visible small differences in the retardation imposed on light waves as they pass through a specimen. The amount of such retardation is proportional to the thickness of the object and the difference between its refractive index and that of the surrounding medium. In order to obtain a good phase-contrast image, it is essential that the phase annuli of the condenser and objective should be accurately centred; this requires repeated checking. The preparation should be as thin as possible; for work at the highest resolution it should

3

not be more than a few microns thick. Preparations of uneven thickness are unsatisfactory. Thin sections (1μm or less) of material embedded in epoxy-resin etc. (p. 65) can often be examined very satisfactorily by phase contrast.

Objects visible in phase-contrast are invariably surrounded by haloes. It may sometimes be advantageous to reduce these by raising the refractive index of the medium (e.g. by using 20% bovine serum albumin). When the refractive index of the medium and object are identical the object becomes invisible. Thus, it is possible to use the phase-contrast microscope to measure the refractive index of cells, and hence obtain information about the concentration of solids they contain (Ross, 1961).

Interference microscopes

Like the phase-contrast microscope, the interference microscope makes visible the relative retardation of light waves passing through transparent objects. The interference microscope, however, is primarily designed for *measurements* of retardation, and has its use in quantitative studies, rather than observation of structure. Knowing the thickness of an object (such as a cell or organelle) and the refractive index of the medium in which it is immersed, it is possible to measure its refractive index and from this determine the concentration of solids it contains and its dry mass. Interference microscopy has applications in studies of both living cells and sections. For accounts of theory and applications see Hale (1958) and Ross (1967).

The *Nomarski interference contrast system* is a special type of interference microscope which, because the image is not greatly affected by structures out of the plane of focus, is particularly valuable for studying the structure of thick objects. For this it has considerable advantages over the phase-contrast microscope. The image is also free from haloes around refractile objects. The theory is described in Padawer (1968).

Polarising microscopy

The polarising microscope is used to detect and measure birefringence. At its simplest it consists of no more than a pair of polaroids mounted on an ordinary microscope, one (polariser) placed below the condenser and the other (analyser) between objective and ocular. With polariser and analyser crossed (i.e. oriented so that they transmit light vibrating in directions at right angles), birefringent objects appear bright on a dark background, if themselves suitably oriented. A simple polarising microscope such as this may be used for morphological studies of strongly birefringent objects (e.g. for studying the distribution of small striated muscles in insects and other arthropods). More elaborate polarising microscopes are used in ultrastructural studies.

The birefringence of biological objects is often *form birefringence*, that is, it arises from the presence of ordered arrays of aniso-diametric objects (fibrils, lamellae etc.) immersed in a medium of different refractive index. Form birefringence can be abolished by infiltrating the specimen with a medium of the same refractive index as the objects. Many substances, however, have an *intrinsic birefringence* which cannot be abolished by such means. It arises from their molecular structure. Measurements of the sign and magnitude of birefringence can give important information about submicroscopic and molecular structure. For the theory and use of the polarising microscope see Bennett (1950).

Fluorescence microscopy

This depends on the fact that some substances fluoresce, i.e. emit light of a different wavelength from that used to irradiate them. The substances commonly studied by this method emit light of visible wavelengths when irradiated with ultraviolet or short wave-length blue light. The method can be used both for detecting intrinsically fluorescent substances (e.g. vitamins A_1 and A_2, ribo-flavin, chlorophyll) and for studying the distribution of fluorescent dyes (fluorchromes), which may be taken up specifically by par-ticular cell components (e.g. acridine orange fluoresces green in combination with DNA, yellowish-green with RNA). Very low concentrations of fluorescent materials can be detected and the method may be the basis of highly sensitive cytochemical tech-niques. In the labelled antibody method (Coons, 1958; Nairn, 1969; Pearse, 1968) antibodies are coupled with a fluorescent

5

dye and then used to stain the corresponding antigen in cells or tissues.

For critical work it is necessary to use a high-pressure mercury arc lamp and a condenser which transmits the exciting radiation freely. Special precautions are necessary to reduce stray light. For details see Price & Schwartz (1956), West (1969), or Pearse (1968). Note that ultraviolet irradiation damages eyes.

Quantitative methods

Dimensions and numbers

The diameter of the field of view given by different combinations of lenses can be measured or calculated and is often useful to know. The *linear dimensions* of objects may be obtained with an eyepiece micrometer. This may be calibrated against a micrometer slide or, for higher magnifications, against a diffraction grating replica.

The *area* of objects can be obtained from photographs or drawings, using a planimeter. Alternatively, squared paper can be used or the relevant parts can be cut out of the photograph or drawing and weighed, the area being obtained from a knowledge of the average weight of the paper per unit area. The *area* and *volume* of particles in sections (e.g. nuclei, secretion granules) can be estimated by random sampling methods which require the use of special eyepiece discs (see Curtis, 1960; Hally, 1964). For determination of the *number* of particles in sections see Eränkö (1955). For the theory and practice of quantitative morphological methods generally, see Weibel & Elias (1967).

The *number of particles in a suspension* (e.g. erythrocytes, bacteria) is most commonly determined by use of a haemocytometer. Accuracy depends on the homogeneity of the sample and accuracy of dilution. The depth of fluid in the haemocytometer must be uniform and this condition is obtained only if the coverslip is in sufficiently good contact with the slide at the sides of the counting chamber for Newton's rings to be present. Samples should be diluted so that 50–100 particles are present in each large square; the number of particles present in the smaller squares into which the large ones are divided should not be more than about 10. A less accurate method is to pipette a small, known volume of the suspension onto a slide and cover with a coverslip of known area. The drop is assumed to spread evenly and counts of particles are made in areas of known size (e.g. using an eyepiece squared graticule).

Many kinds of particle in suspension (e.g. bacteria, erythrocytes, protozoa, isolated cells, isolated organelles) can be rapidly counted and sized automatically with an electric sensing device (e.g. Coulter counter). In this the particles are made to pass through a small aperture and the change in conductivity of the liquid in the aperture associated with the passage of each particle is measured and recorded. The principles and apparatus are discussed in Kubitschek (1969) and Harvey (1968).

Microspectrophotometry

This method permits the estimation of certain substances present in selected areas of tissue sections or single cells. For an account of the method and its uses see Pollister, Swift & Rasch (1969), Freed (1969) and Ruthmann (1970). The principle is the same as that used in the spectrophotometric estimation of substances in solution: the proportion of light absorbed by a solution is dependent on the concentration of the absorbing substance. The basic law of light absorption, the Lambert–Beer Law, states that, if I_0 is the intensity of the incident light and I_s the intensity of the light after passage through the specimen,

$$\log \frac{I_0}{I_s} = kcd$$

where c is the concentration of the absorbing substance, d the thickness of the absorbing layer, and k the extinction coefficient. The latter is a function of the wavelength.

Measurements of absorption in the u.v. can be made on unfixed, unstained cells; in particular, absorption at 260 nm can be used to measure nucleic acids. At visible wavelengths the method is applied to material stained by standard histochemical methods. Microspectrophotometry has been used extensively for estimation of DNA in Feulgen-stained sections. Both DNA and RNA can also be measured following staining with azure A or other stains, DNase or RNase being used to provide specificity. Proteins can be estimated in sections stained with naphthol yellow, mercury brom-phenol blue or the Millon reaction. It is possible to make measurements on areas down to 0.3 μm in diameter. The DNA content of single bands in polytene chromosomes has been measured.

Ruthmann (1970) describes a relatively simple microspectro-photometer for measurements at visible wavelengths, which can be assembled from commercially available components. More complex instruments are made by some optical firms. For precise measurements on small areas or of weak absorption, considerable care is needed to achieve the necessary stability of amplification.

An important limitation of the technique arises from the fact that the Lambert–Beer Law applies to absorption in solutions and does not necessarily hold for absorption in fixed and stained sections. The extinction coefficient (k) may be substantially different in a tissue section from what it is in solution. For this and other reasons it is frequently possible to obtain estimates only of the relative amounts of substances, rather than of their absolute concentrations. For a discussion of sources of error see Ruthmann (1970).

Methods of recording

Drawing

Accurate drawings of sections can be made rapidly by using a squared eyepiece graticule (0.5 mm squares) and squared paper. Alternatively, and especially for details, a *camera lucida* may be used. With this the observer sees the image of the specimen superimposed on the drawing paper and traces the outlines with a pencil. It is essential to match the intensities of illumination in the microscope and on the paper. Accurate drawings can also be made by direct projection, using a high-intensity light source and a mirror or projection prism mounted over the eyepiece. A darkened room is necessary. Note that with all methods linear dimensions may be accurately reproduced only in the centre of the field of view.

For the technique of making drawings for reproduction see Cannon (1936) and Staniland (1952).

Serial reconstruction

See Pantin (1948). For organs extending over many sections (e.g. the nervous system of an annelid) it is best to make the reconstruction along an axis at right angles to the plane of the sections. Use a squared eyepiece graticule and orient this so that the centre line runs through some axis or reference point that can be identified in each section. Starting with the first section, measure the position of the organ concerned on each side of the axis. Draw a line on squared paper to represent the axis and transfer the measured distances to one of the transverse lines on the paper (figure 1). Repeat for successive sections. Choose suitable scales so that transverse and longitudinal distances correspond: the section thickness must be known.

Photomicrography

See Allen (1958). For the theory and practice of photography generally see James (1966) and Engel (1968). The only essential apparatus required is a camera body of some sort to support the plates or film, and some means of focusing the image and making the exposure. No additional optics are necessary. Exposures are

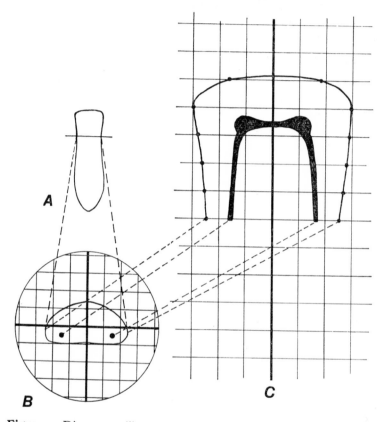

Figure 1. Diagram to illustrate a method of reconstruction from serial sections. *A* is the specimen; *B* is a section at the level indicated, viewed with a squared eyepiece grid and positioned symmetrically about the centre line of the grid; *C* shows how distances measured on the section are transferred to squared paper, the lines on which correspond to those on the eyepiece grid. The process is repeated for successive sections. (After Pantin, 1948.)

determined either by making trial exposures (e.g. 1, 2, 4, 8, 15, 30, 60 s) for each set of optical conditions, or by using a photoelectric meter. For photography of moving objects an electronic flash is necessary. The contrast and definition of micrographs may be substantially improved by the use of colour filters. To increase contrast use filters of a colour complementary to that of the stain (e.g. use a green filter for Feulgen-stained nuclei). Neutral grey filters may be used to adjust light intensity without altering the spectral composition of the light. The choice of emulsion is determined by factors similar to those operative in ordinary photography. Films such as Ilford 'Micro-neg' or Kodak 'Microfile pan' are fine-grained and of excellent contrast but are slow. Emulsions such as Ilford 'Pan-F' or Kodak 'Pan-X' are faster, give adequate contrast and are more commonly used.

Living material

General notes

Always examine material alive if possible. Rapidly moving organisms such as protozoa can be narcotised (p. 84) or put in methyl cellulose (p. 84). Tissues should be teased apart with needles before mounting lest they squash under the coverslip or are too thick for observation. See p. 18 for other methods of separating living tissues into their component cells.

Do not mount tissues in too much fluid when using oil-immersion objectives, otherwise the coverslip will stick to the objective and the specimen will move as the microscope is focused.

Tissues mounted under a coverslip will become anoxic; this can be avoided by mounting in a *hanging drop*. Place tissues in a *small* drop of medium spread thinly in the centre of a large coverslip. Smear Vaseline around the edge of the cavity of a large cavity slide, and invert this over the coverslip so that the cavity is over the drop. Make sure the sides of the cavity do not touch the drop and that the Vaseline forms a seal. Turn the slide over carefully. This method of mounting has the disadvantage that the water surface is curved, so continual centring of the substage optics is necessary, and droplets of moisture condense on the cavity of the slide. Slightly warming the slide may help. These difficulties are avoided with *perfusion slides*, in which the specimen is overlaid with dialysis cellophane (Rose, 1967). Many samples of cellophane dialysis tubing, however, release toxic substances.

When examining living material always use an effective heat filter (glass ones are most convenient).

Living material is commonly transparent. Its structure can sometimes be shown up by reducing the aperture of the condenser iris, but this reduces the resolving power of the microscope. *Phase-contrast microscopy* (p. 3) is better.

Vital stains

These are not now much used. *Janus green* will stain mitochondria in living cells. Use a very dilute solution (see p. 84) in an appropriate saline medium (p. 104). *Neutral red* as a pale orange solution in saline will stain living macrophages, but not living fibroblasts. Neutral red and phenol red are *colour indicators* of the pH of living tissue. Sulphonphthalein stains injected into the cytoplasm give a more reliable measure of the pH of cytoplasm (Chambers & Chambers, 1961).

Tissue culture

The practical details of most methods for animal cells are given in Paul (1970), and for plant cells in Gautheret (1959).

Manipulation of living material

Tungsten instruments are useful for fine dissection. They are harder than steel, can be heated in a flame (for cleaning and sterilising) without losing their hardness and can be sharpened electrolytically. Fine tungsten wire is brazed by means of the silver brazing alloy Easy-flo No. 2 (Johnson, Matthey & Co. Ltd, Hatton Garden, London EC1) onto suitable holders (e.g. metal forceps). The wire is dipped into brazing alloy which is kept molten in a steel crucible by heating with an oxy-gas flame. The surface of the Easy-flo must be covered with a layer of clean Easy-flo flux. Make the end of the tungsten wire red hot, then push it through the flux into the Easy-flo. After 10 s remove the wire quickly, so that it does not become red hot. Repeat this process on the tip of the holder, then bring tungsten wire and holder together and heat until the brazing alloy just melts. Boil the instrument in water to remove the flux. Instruments are bent to shape, then sharpened electrolytically in 10% KOH, using 2–6 volts A.C. The rate of dissolution varies with the voltage and the distance between the electrodes. Dissolution is most rapid at the tip of the wire, which soon acquires a steeply tapering point. A parallel-sided needle may be produced by immersing a long length of the wire for long periods; a tapering needle is produced by rhythmically moving the tip of the needle in and out of the KOH. To make forceps, mount the tungsten tips vertically in the bath so that both are just below the surface. Electrolysis will bring them both to the same length at the surface of the liquid. They are then sharpened as described above (watch under a microscope). The flat surface where one tip bites against the other is produced by folding a piece of polishing paper with its rough surfaces outwards, gripping it between the tips and moving it to and fro a few times.

A simple type of *micromanipulator*, useful for holding or placing instruments accurately, has been described by Goldacre (1954). It is made from microscope slides. Heavy instruments are better mounted on a screw type of instrument (Chambers or Pryor). Micromanipulators operated by a joy-stick (Leitz, Singer) are particularly useful for rapid, complex movements. Instruments for microdissection are made from the splintered edge of a razor blade, from selected splinters of glass, or on a de Fonbrune micro-forge (Beaudouin et Cie, 1 & 3 Rue Rataud, Paris V^e; English agents: Scientific Techniques Ltd, Brockham, Betchworth, Surrey). The latter is particularly useful for constructing micro-pipettes of precise size and shape for injection or for holding cells and tissues by suction.

Localised lesions in cells can be produced with an *ultraviolet microbeam* or *laser light* (see Moreno, Lutz & Bessis, 1969).

Narcotisation

This is best avoided if the tissues are to be examined cytologically. For physiological and other experiments keep aquatic animals in good conditions in a cool, quiet place for several hours so they relax, then increase the concentration of narcotic around them drop by drop. This may take 30 min to several hours. Stop adding narcotic if animals show signs of agitation or convulsion. The following methods are useful (see Pantin, 1948).

10% ethanol. Make up from absolute ethanol, rather than industrial spirit. Use for freshwater invertebrates.

$MgCl_2$ *solution.* For marine animals use a 7.5% solution of $MgCl_2.6H_2O$, diluted with an equal volume of sea water. When superficial narcosis has occurred after a few minutes the process may be hastened by injection of the solution internally.

Menthol. Scatter some crystals on the surface of the water and leave overnight. Good for sessile marine and freshwater animals (e.g. Polyzoa).

Ether vapour. Good for insects, arachnids and terrestrial vertebrates.

MS 222. See the manufacturer's information (Technical Bulletin No. 1, Sandoz Products Ltd, 41 Upper Grosvenor St, London W1X 0AL). Excellent for fish and amphibia. Immerse the animal in a fresh aqueous solution of 1 part MS 222 in 12500 to 25000 parts water. Fish should be immobilized in 2-4 min. Bell (1964) tabulates the properties of this and other anaesthetics for fish.

CO_2. This may be applied in the form of soda water for freshwater animals.

Low temperature. For invertebrates keeping the animal in a refrigerator often gives adequate anaesthesia.

Drowning. Many insects, especially Hemiptera, can be anaesthetised by drowning. Insert the animals under an inverted beaker completely filled with tap water, standing in a larger dish of water. Leave until they have stopped moving for a few minutes.

Separation of tissues into cells

Living animal tissues

These can be dissociated by the following methods, used separately or together.

Trypsin

> This is the most effective method for most tissues.
> (1) Incubate tissues in a calcium- and magnesium-free saline, 15 min.
> (2) Transfer to a fresh solution of the saline containing 1% crystalline trypsin.
> (3) Incubate for 20 min at 38 °C or 1 h at room temperature.
> (4) Rinse the tissues *gently* with calcium- and magnesium-free saline.
> (5) Flood with the medium in which the cells are to be dispersed.
> (6) Dissociate by sucking tissues briskly into and out of a pipette.

Trypsin adsorbed to cells impairs their viability. It can be destroyed by treating them with a 1% solution of soybean trypsin inhibitor in the saline for 10 min after stage (3).

Collagenase

See Lasfargues (1957). Even the 'purified' enzyme may have toxic contaminants.

Mechanical disruption

Loosely aggregated tissues or those treated with enzymes can be forced through bolting silk. The method has the disadvantages that large quantities of tissue are needed, cells are not fully disaggregated, and all the cells may be affected by those damaged mechanically.

Chelating agents

This is effective only on early embryos (Zwilling, 1954). Tissues of adults cannot be dissociated with these agents without causing extensive damage, both mechanical and metabolic (Moscona, Trowell & Willmer, 1965).

Ultrasound

This has been used to disperse mammalian cells (Moore, Williams & Sanders, 1971).

Fixed animal tissues

Maceration is used to investigate the shapes of cells in tissues; living cells from animals tend to round up when isolated.

The Hertwigs' method (1879)

This is good for preserving the shape and detailed cytology of cells.

(1) Fix small pieces of tissue in a mixture of 0.2% acetic acid and 0.04% osmium tetroxide in a suitable saline (p. 104).
(2) Transfer tissues to 0.2% acetic acid for several days.

Tissues should fall apart when teased gently with needles.

Goodrich's method (1942)

Prepare a saturated solution of boric acid in calcium- and magnesium-free saline. Add 2 drops of Lugol's iodine solution (p. 84) to every 25 ml. Immerse the tissue in this for several days. It can then be disrupted mechanically. The cells will keep in this solution for a few weeks, but are not as well preserved as by the Hertwigs' method.

Fresh plant tissues

These can be macerated by soaking them in snail digestive juice for 15–45 min. Use the juices from the stomachs of snails collected in early afternoon. Alternatively, snail digestive juice can be bought (Koch-Light). See p. 76 for another method of macerating plant tissue.

Fixation and fixatives

General

The object of fixation is to preserve the structure of cells and tissues in as lifelike a manner as possible, and to do this in such a way that they may be subjected to further procedures such as sectioning and staining. Most fixatives act by precipitating or cross-linking proteins, so that a stable three-dimensional network is formed. Autolysis and bacterial attack are inhibited and staining properties may be improved. The chemistry of fixation is not entirely understood, but it is known that different fixatives react with cell components in a variety of ways (see Baker, 1958; Pearse, 1968). Some cell components are not rendered insoluble by the commonly used fixatives. Neutral fats, for example, usually remain unfixed and are extracted if material is treated with fat solvents during subsequent processing. Substances of low molecular weight (ions, sugars etc.) are commonly removed. On the whole, only the macromolecular skeleton is preserved.

There is no single ideal fixative for all purposes. Most of the commonly used ones are complex mixtures which have been found empirically to give satisfactory results for particular purposes. For *cytological* work avoid fixatives which cause gross precipitation of proteins; aldehydes (especially formaldehyde and glutaraldehyde), potassium dichromate and osmium tetroxide are most commonly used. For *histology* more drastic fixatives are usually preferable, since they leave tissues in better condition for sectioning and staining; the destruction of fine cytological detail which they may cause is not important.

Fixatives penetrate tissues at different rates and react at different speeds. Where possible fix small pieces of tissue, preferably not more than a few mm thick. With slowly penetrating fixatives, such as osmium tetroxide, even smaller pieces are necessary. Hollow organs (intestine etc.) should be opened if possible to allow free access of fixative. Material which has to be dissected out (e.g. pieces of mammalian organs) should be placed in fixative as soon as possible, in order to minimise post-mortem changes. For perfusion methods see Sjöstrand (1967). There are few rules with regard to the duration of fixation; optimum times must be found by experiment, and the suggested times below are intended only as a guide. The volume of fixative used should be 15–20 times that of the tissue, at least.

Fixatives for animal tissues

A great many fixative mixtures for animal histology have been described; details can be found in the standard works (Lillie, 1965; Drury & Wallington, 1967). The ones described below will be found adequate for most purposes. For special tissues (blood, nervous systems) see the relevant sections.

For histology fixatives are usually used at room temperature. Warm fixatives (e.g. Bouin at 30–60 °C) may be useful for organisms which contract rapidly on fixation.

See that animals are free from grit before fixation; it causes damage to paraffin sections. If possible, starve animals for a day or so to clear the gut.

Formaldehyde

This is commonly used as a fixative for routine histology and pathology. It is usually obtained as formalin, which is a 40% solution of formaldehyde in water. This is diluted to a 4% solution (= 10% formalin) for use as fixative. Store formaldehyde solutions over calcium carbonate or Amberlite IR-45 resin, to neutralise acidity, or use buffered, as recommended by Lillie (1965):

Formalin (40% HCHO)	100 ml
Distilled water	900 ml
$NaH_2PO_4.H_2O$	4 g
Na_2HPO_4	6.5 g

Very pure formaldehyde solutions, preferable for some purposes (Burgos, Vitale-Calpe & Téllez de Iñon, 1967), can be prepared by dissolving the appropriate amount of *paraformaldehyde* powder in water at 60–65 °C and adding a few drops of concentrated NaOH until the solution clears (pH 7.1–7.3). Alternatively, dissolve the paraformaldehyde first in the alkaline component of a buffer at

60–65 °C and then add the acid component. Solutions prepared from paraformaldehyde should be used fresh.

Formol-saline is 4% formaldehyde containing 0.75% NaCl; the salt may help to reduce distortion resulting from osmotic changes. For marine organisms make up in sea water. For preservation of lipids, especially phospholipids, use *formaldehyde–calcium*: this is 4% formaldehyde containing 1% calcium chloride (anhydrous), stored over calcium carbonate (Baker, 1944).

In all cases fix for 16–48 h (tissues may be stored in the fixative). Wash in several changes of 50% or 70% ethanol (12 h in all), unless lipids are to be preserved, in which case see p. 53.

Susa

Heidenhain's Susa is particularly good for invertebrate histology.

$HgCl_2$	45 g
NaCl	5 g
Distilled water	800 ml
Trichloracetic acid	20 g
Acetic acid (glacial)	40 ml
Formalin (40% HCHO)	200 ml

The first three components can be kept as a stock solution and the others added before use. For marine animals it may be advantageous to raise the NaCl to 30 g or more, or to use sea water instead of distilled water. Fix for 3–24 h, then transfer directly to 96% ethanol and wash for 12 h.

Fixatives containing mercuric chloride may give rise to black precipitates in fixed material. Susa does not usually cause these but it is sometimes recommended that the alcohol used for washing should have iodine (0.25–0.5%) added to it.

Zenker

This is an excellent general fixative and may preserve cytological detail better than other histological fixatives.

$HgCl_2$	5 g
$K_2Cr_2O_7$	2.5 g
$Na_2SO_4.H_2O$	1 g
Distilled water	100 ml
Acetic acid (glacial)	5 ml

The first four components are kept as a stock solution; the acetic acid is usually added before use (it can be omitted altogether, or replaced with an equal volume of formalin to give *Helly's fluid*, which is an excellent cytoplasmic fixative). Fix for 3–18 h. Wash overnight in running tap water, then transfer to 50% or 70% ethanol.

Bouin

Bouin is good for marine invertebrates and for most mammalian tissues (except kidney, and organs containing many mucous cells).

Picric acid (saturated aqueous solution)	75 ml
Formalin (40% HCHO)	25 ml
Acetic acid (glacial)	5 ml

The solution keeps indefinitely. Fix for 12 h or longer (material may be stored in Bouin). Wash in several changes of 70% ethanol, 24 h.

Alcoholic Bouin (= Duboscq–Brasil)

This is a highly penetrating fixative, useful for arthropods etc. with impermeable cuticles.

Picric acid	1 g
Acetic acid (glacial)	15 ml
Formalin (40% HCHO)	60 ml
Ethanol (80%)	150 ml

Fix for 2 h or more. Wash in several changes of 90% ethanol, 24 h.

Fixatives for plant tissues

See Johansen (1940); Sass (1958); Purvis, Collier & Walls (1966). The tendency of some plant materials to float when placed in fixative, because of entrapped air, can be overcome by reducing the pressure with a filter pump.

Formalin-aceto-alcohol (FAA)

This is the most commonly used fixative for plant material.

Ethanol (50 or 70%)	90 ml
Acetic acid (glacial)	5 ml
Formalin (40% HCHO)	5 ml

Fix for at least 12 h. Material can be left in the fixative indefinitely. Wash in 50% ethanol.

Chrom-acetic mixtures

	weak	medium
Chromic acid (10% aqueous)	2.5 ml	7 ml
Acetic acid (10% aqueous)	5.0 ml	10 ml
Distilled water	92.5 ml	83 ml

Of these mixtures, weak chrom-acetic is used for more delicate materials (algae, fungi, bryophytes etc.) and medium chrom-acetic for root-tips, ovaries, ovules, etc. Fix for 24 h or more; wash thoroughly in running water.

Chrom-acetic–formalin mixtures

These, termed Navashin fluids, are used principally for cytological preparations, such as smears of anthers, squashes of root-tips etc. Many variants exist, of which the following is perhaps the most useful.

A.	Chromic acid	1 g
	Acetic acid (glacial)	7 ml
	Distilled water	92 ml
B.	Formalin (40% HCHO)	30 ml
	Distilled water	70 ml

Mix equal parts of A and B before use. Fix for 12–24 h and wash in 70% ethanol.

Bouin (p. 23)

This is useful for plant as well as animal material and is good for root-tips, embryo sacs, etc.

For cytological studies *Flemming*-type mixtures (p. 25) can be used. For details see Sass (1958).

Fixatives for cells

For fixation of nuclei and chromosomes see p. 76; for protozoa p. 85; for electron microscopy p. 60.

The fixatives used for electron microscopy, principally osmium tetroxide and glutaraldehyde, can also be used for light-microscope studies and are recommended when the best possible preservation is required. Excellent results are obtained if their use is followed by embedding by the agar/ester wax method (p. 32) or in methacrylate (p. 35) or epoxy-resin (p. 61). Paraffin embedding is not recommended.

Of the histological fixatives described above, formaldehyde, Bouin, Helly or Zenker may give adequate preservation of cytological detail for some purposes. For better preservation of cytoplasmic structure one of the numerous osmium-containing mixtures may be used; note, however, that few stains except iron haematoxylin and some basic dyes can be used satisfactorily after these.

'Flemming-without-acetic'

This is a useful osmium-containing mixture:

Osmium tetroxide (2 %)	4 ml
Chromic acid (1 %)	15 ml

This should be made up as required. Sodium chloride may be added to make it isotonic (e.g. 3 % for marine organisms, 0.75 % for others). Fix small pieces of tissue for 24 h, then wash in running tap water, 2–5 h. Transfer to 30 % ethanol and dehydrate slowly.

Champy's fluid

This is another osmium-containing mixture, excellent for protozoa and the cytoplasm of cells generally.

Chromic acid (1 %)	7 ml
Potassium dichromate (3 %)	7 ml
Osmium tetroxide (2 %)	4 ml

Fix and wash as for Flemming.

Smears of cells, or cells growing on coverslips, may be fixed with *osmium tetroxide vapour* by placing for 1 min in a closed vessel containing a 2 % solution. This may be followed by fixation in an osmium-containing fixative such as Flemming-without-acetic or Champy.

Freeze drying and freeze substitution

Freeze drying

See Pearse (1968). In this process the tissue is rapidly frozen (quenched), then dehydrated in the frozen state by subliming off the water in a vacuum. It is an excellent method of preparing specimens for many purposes. Since the method does not result in fixation, enzymes are kept in an active state for enzyme histochemistry. Unfixed paraffin sections of such material can be cut. Even soluble substances of low molecular weight should be kept in their correct position, provided quenching is fast enough and drying performed at a low enough temperature (at least $-60\,°C$ for some substances).

Quenching

Small pieces of tissue are placed on small squares of aluminium foil and plunged rapidly into a tube of isopentane or Freon 22 (obtainable from I.C.I. Ltd) cooled to $-165\,°C$. The quenching liquid is contained in a 7.5×2.5 cm specimen tube inserted in the neck of a vacuum flask filled with liquid nitrogen. Tissues are frozen within a few seconds but need not be removed immediately.

Drying

The frozen specimens are rapidly transferred to a suitable apparatus (Pearse, 1968). The Edwards–Pearse tissue dryer (Edwards High Vacuum Ltd, Crawley, Sussex) is convenient. For histological purposes the tissue must be kept at -40 to $-60\,°C$ during drying, otherwise large ice crystals will grow and disrupt the structure. Drying may take up to 3 days. Tissues may then be embedded in paraffin wax, without further treatment, or can first be fixed to stabilise them during further processing. Tissues can be made slightly insoluble by immersing them in absolute ethanol. Under these conditions the alcohol does not fix tissues. More fixation can be produced by immersing the dried tissues in 90% ethanol.

Formaldehyde vapour prepared from paraformaldehyde and used at 60 °C for 1–3 h is an excellent fixative of freeze-dried material to be used for standard histochemistry.

Freeze substitution

This method, in which quenched tissues are transferred to a cold dehydrating agent, has many of the advantages of freeze drying but does not need expensive drying apparatus and the specimen does not have a vapour phase passing through it. A disadvantage is that lipids dissolve in the drying agent. Quenched specimens are transferred to a tube of acetone or ethanol at − 70 °C containing the drying agent Molecular Sieve 4A (Union Carbide, Linde, obtainable from B.D.H. Ltd, Poole, Dorset). The cooling can be achieved with a solid CO_2 and acetone mixture. Dry for 2–7 days with intermittent stirring. Allow specimens to warm up to 0 °C in the dehydrating agent, then transfer to fresh absolute ethanol at this temperature. The specimens can then be cleared and infiltrated with paraffin wax. For further details see Pearse (1968).

Sectioning methods

General notes

This chapter deals with sectioning methods for the light microscope; for electron-microscope methods see p. 59.

Some plant tissues can be sectioned freehand if merely supported externally (e.g. in elder pith or cork). In general, however, tissues require better support and this involves impregnating them with some substance which sets hard in and around them and can then be sectioned. Embedding media are for the most part of two kinds: those which are liquefied by melting and harden on cooling, and those which are liquid in the monomeric state and harden on polymerising. Paraffin and other waxes belong to the first category, methacrylates and other resins to the second.

Material to be sectioned is usually first fixed and washed. Then, since the common embedding media are not miscible with water, it is usually necessary to dehydrate, and to follow dehydration by soaking in a solvent which is miscible with both dehydrating agent and embedding medium. This is followed by infiltration with the embedding medium in liquid form, which is then hardened. The block is sectioned on a microtome, the sections mounted on slides, the embedding medium dissolved out and the sections stained. Numerous variations on this basic procedure are possible. In general, for most histological and histochemical purposes paraffin embedding will be satisfactory. The special advantages of other methods are indicated below.

Paraffin

This is simple to use, has good sectioning and ribboning properties and embedded material can be stored indefinitely. Sections down to 1 μm can be cut. Its chief disadvantage is that material has to be completely dehydrated before embedding: most lipids are removed and there may be considerable shrinkage.

For *dehydration* ethanol is most frequently used. Material is taken through a series of solutions of increasing concentration, usually 70%, 90%, 95% and 100%. For average-sized blocks of tissue suitable times are 3–12 h in the lower grades of ethanol and 3–8 h in 100%, which is changed two or three times. For large or delicate objects additional grades of ethanol may be inserted (e.g. dehydration is started in 50%) and times may be extended. It is not certain, however, that such gradual dehydration results in any considerable improvement in preservation. For small pieces of tissue dehydration may be speeded up (e.g. 1–3 h in each alcohol). There is in general no disadvantage in extending the times in the lower alcohols, but tissues should not be kept in 100% ethanol longer than necessary, otherwise they may harden. Material which was washed in alcohol after fixation is simply transferred to the next stronger grade of ethanol and carried on from there. *Thorough dehydration is important.* Inadequate dehydration is the commonest cause of poor results. Make sure that alcohol (or other dehydrating agent) has not picked up water. Keep bottles, vials etc. tightly closed.

Other common dehydrating agents include acetone, dioxane, Cellosolve (ethylene glycol monoethyl ether) and *n*-butanol. Of these, Cellosolve is a less violent dehydrating agent than ethanol. Transfer material from water to 50% Cellosolve and then to 100% Cellosolve. It causes less shrinkage and hardening than ethanol but extracts more lipid and other constituents. For further information on dehydration see Lillie (1965) and Baker (1960).

Tissues are taken from the dehydrating agent into a paraffin solvent ('clearing agent' or 'antemedium'). Toluene or xylene are commonly used. Delicate specimens may be taken first into a 1:1 mixture of one of these with the dehydrating agent for 15–60 min and then into pure solvent. Clear for 1–8 h (keep the time to a minimum to prevent brittleness). After this material is transferred to molten paraffin. For yolky specimens or where absolute alcohol must be avoided, use *methyl benzoate* as clearing agent. Material may be transferred to this from 95% ethanol. Layer the methyl benzoate below the alcohol: the specimens first float at the interface, then sink. Pipette off the alcohol. Change the methyl benzoate two or three times, then transfer to paraffin via benzene (p. viii), since paraffin is not freely miscible with methyl benzoate.

For most purposes paraffin of melting point 52–54 °C is suitable.

Harder wax is necessary for thinner sections or for sectioning in warm conditions (for the latter see Gray, 1954). Wax should be kept molten at a temperature a few degrees above its melting point. Transfer specimens to it with the minimum of solvent. There is probably no advantage in using an intermediate bath of paraffin in solvent, as is sometimes recommended. Infiltration with molten wax usually takes 1–3 h in all and two or three changes of paraffin should be used. Do not leave material in molten paraffin longer than necessary: it becomes brittle. Specimens are transferred from one bath to the next with a warm spatula. The paraffin can be used repeatedly. A schedule for dehydration and embedding in paraffin is given in table 1.

TABLE 1. *Schedule for embedding in paraffin wax*

(The times given are for specimens 3–5 mm thick.
For smaller objects they may be reduced.)

(1)	Fix and wash	
(2)	70 % ethanol	3–12 h
(3)	90 % ethanol	3–12 h
(4)	95 % ethanol	3–12 h
(5)	100 % ethanol I	1–2 h
(6)	100 % ethanol II	1–3 h
(7)	100 % ethanol III	1–3 h
(8)	Toluene (or xylene)	1–8 h
(9)	Paraffin wax, 3 changes	1 h each
(10)	Embed in fresh wax	

Embedding is done in a solid watchglass, in paper or foil boats or in a mould made from brass L-pieces. Glass or metal moulds should be *thinly* smeared with glycerol to prevent sticking. Fill the vessel with fresh molten paraffin, add the specimen and orient if necessary with a warm needle. Then cool: the surface layer should be allowed to solidify before sliding the block vertically on edge into water. Rapid cooling is said to give blocks of smaller crystalline structure and better cutting properties than does slow cooling. Small or transparent specimens are more easily seen in the block if stained with 1 % eosin in 95 % ethanol during dehydration.

For *sectioning*, cut out a block of wax containing the material and seal it to the block holder of the microtome by melting the wax with a hot scalpel. Trim the block to a four-sided truncate pyramid and make sure that the faces which are to lie parallel to the knife edge are accurately parallel. This ensures that ribbons will be straight. If it is important to keep track of the orientation of the sections, trim one of the lateral faces of the block obliquely. Trimming should be done with clean strokes of a sharp scalpel or razor blade. Irregular block faces may cause irregular or broken ribbons.

For microtomes, microtome knives and knife sharpening see Drury & Wallington (1967). In sectioning make sure that the faces of the block trimmed parallel to each other are accurately aligned with the knife edge. Cut at 8–10 μm for routine purposes. Use a moist paint brush for handling ribbons and place them in series on paper as they are cut.

Faults in sections are dealt with by Steedman (1960) or Drury & Wallington (1967). A sharp knife is essential for good sectioning. Make sure that knife and block are securely clamped. Scratched or split ribbons result from a damaged knife or grit in the block etc. Curved ribbons usually arise from inaccurate trimming of the block faces, or occasionally because of the presence of a local source of heat, such as a lamp, causing uneven expansion. Curling up of sections is commonly caused by a blunt knife or by a knife set at too large an angle to the perpendicular; sometimes, if the first section is cut slowly and held flat with a paint brush while being cut, the successive sections will also come off flat. Static electricity may cause ribbons to be very difficult to handle; it may help to breathe on the block and ribbon.

Mount sections on clean, grease-free slides (p. 116). As adhesive, Mayer's albumen is traditionally used. This is prepared by mixing egg white with an equal volume of glycerol; filter through coarse filter paper and add a few crystals of thymol or other preservative. A *small* drop of albumen is rubbed evenly over the slide with a clean finger. It is important that the layer should be as thin as possible, otherwise sections may wash off and the albumen may take up stain. Cut ribbons into suitable lengths (not more than 50 mm long for a 75 mm slide), and place these on the slide in order, dull side up and starting with the first section in the top left-hand corner. Number the slide at the bottom right-hand corner with a diamond marker. Then flood the slide gently with distilled water from a pasteur pipette or wash bottle, so that the sections float, and place on a hotplate at a temperature a little below the melting point of the wax. The sections will extend as they warm up. When they are fully stretched, arrange them in their final positions with a needle and drain off the water. Dry overnight at about 30 °C (e.g. on top of the paraffin oven).

Before staining, dissolve the paraffin in xylene (1–3 min), then take through 100% and 70% ethanol (30–60 s in each) to distilled water.

Ester wax

This is based on diethylene glycol distearate. The original formula was later modified and the new mixture is called ester wax 1960 (Steedman, 1960). In spite of its low melting point (45–47 °C) the wax is hard and sections down to 1 μm can be cut with little compression. Good ribbons are formed. Since infiltration is carried out at a relatively low temperature, there is less shrinkage than with paraffin of comparable hardness. Ester wax adheres well to hard, smooth materials, such as insect cuticle. It is obtainable from British Drug Houses Ltd, Poole, Dorset.

Ester wax is soluble in a variety of solvents; xylene, Cellosolve, n-butanol and 95% ethanol are recommended. Material is dehydrated as for paraffin, in ethanol or Cellosolve, then passed through one of the above solvents (3 changes, 8 h in all) and a mixture of wax and solvent (1:1 overnight) before infiltration with wax. Three changes are used and infiltration takes about 4 h for pieces of material up to 5 mm thick. For embedding, fresh wax is heated to 55–60 °C and placed in metal moulds. The tissue is added and the mould surrounded by iced water. The surface of the block should be kept liquid with a hot spatula until the block has solidified all through. Do not immerse the block in water. Sections can be cut on a conventional microtome, provided the knife is reasonably heavy and rigidly mounted. The sections are floated on water on slides coated thinly with Mayer's albumen, stretched at 40–45 °C and dried at the same temperature. Dewaxing and staining are as for paraffin sections. For further details see Steedman (1960).

Wigglesworth (1959) uses ester wax after preliminary embedding in agar for cutting sections in the range 0.5–1 μm. The technique is of great value in cytological work at the light-microscope level.

(1) Fix and wash.
(2) Place in 5% agar at 60 °C for 1 h.
(3) Allow to set at room temperature, if necessary orienting the tissue while the agar is cooling.
(4) Trim the agar block until it is slightly larger than the final block that will be cut.
(5) Pass through 30%, 50% and 70% ethanol, 30 min each.
(6) 70% ethanol plus Cellosolve (2:1), 30 min.
(7) 70% ethanol plus Cellosolve (1:2), 30 min.
(8) Cellosolve (3 changes), 30 min.
(9) Cellosolve plus ester wax (1:1), 30 min.
(10) Ester wax (2 changes at least), 30 min.
(11) Leave overnight in the final ester wax at 55–60 °C.
(12) Embed and cool rapidly.

N.B. Ester wax of Steedman's original (1947) formula must be used; the 1960 formula will not penetrate the agar block. Trim the block so that the agar extends to the surface on all sides. Sections can be cut dry or on to the surface of water. A conventional micro-tome (e.g. Cambridge rocking microtome) with an ordinary knife may be used, but for thin sections (0.5–1 μm) use either sharpened razor blades mounted in a special holder (Wigglesworth, 1959) or an ultramicrotome and glass knives. Sections are picked up on coverslips. If cut dry float out on 20% ethanol. Dry down on the hotplate as for ester wax sections. Wigglesworth (1971) describes the use of a paper barrier for keeping sections together in the microtome trough before they are picked up.

Polyethylene glycol methacrylate

Polyethylene glycols (sold as Carbowax, Polywax, etc.) have been used as an embedding medium instead of paraffin. They are miscible with water and material can be infiltrated and embedded without dehydration. The principal disadvantage is that the wax is soluble in water and most other liquids, giving rise to considerable difficulties in floating out sections on slides. For this reason the method is not now much used (for details see Steedman, 1960; Lillie, 1965).

Polyethylene glycol can, however, be usefully combined with methyl methacrylate to form mixtures that permit large sections to be cut down to 2 μm in thickness, using a conventional microtome. The method is useful for hard objects, such as insect cuticle and bone, and also for soft tissues of high water content. For details see von Hirsch & Boellard (1958) or Ruthmann (1970).

Celloidin

Celloidin is a form of cellulose nitrate. It may be used as an embedding medium for hard or brittle objects, or for large specimens such as whole eyes, or in cases where thick (100 μm) sections are required (e.g. brains). Celloidin embedding causes less shrinkage than does paraffin. Its disadvantages are that infiltration is slow, sections do not form ribbons and cannot be cut thinner than about 10 μm, and there is no reliable method for attaching them to slides. The following procedure is recommended by Drury & Wallington (1967); for celloidin embedding of plant material see Sass (1958).

(1) Fix, wash and dehydrate as for paraffin.
(2) Ethanol (100%) and ether (1:1), 24 h.
(3) Celloidin (2%) in ethanol–ether (1:1), 5–7 days.
(4) Celloidin (4%) in ethanol–ether (1:1), 5–7 days.
(5) Celloidin (8%) in ethanol–ether (1:1), 3–4 days.
(6) Embed in 8% celloidin in ethanol–ether in paper boxes or trays.

Cover tissue blocks to a depth of 30–40 mm and leave a margin of at least 10 mm around the tissue. Place the boxes in an air-tight container (e.g. a desiccator) and place a small dish of ether with them, to promote dispersion of air bubbles. After a few hours replace the ether with a dish of chloroform. Leave until the block has the consistency of hard rubber. Store blocks when hardened in 70% ethanol and do not allow to dry out.

For sectioning, glue the block to the object holder of the microtome with either thick celloidin solution or a commercial adhesive (e.g. Durofix), first softening the under surface of the block by

brief immersion in ethanol–ether. The most suitable kind of microtome is the sliding or base sledge type, with the knife slanted at
about 45° to the direction of movement of the block. A plano-
concave knife is recommended, though not essential. Keep block
and knife wet with 70% ethanol and cut slowly. Transfer sections
to a dish of 70% ethanol with a brush or forceps.

Celloidin sections are best stained loose, with the embedding
medium still present. Staining times may be longer than with
paraffin sections and more dilute stains may be advantageous.
After staining mount as follows.

(1) Take sections to 95% ethanol.
(2) Place sections individually on slides and cover with a piece
    of absorbent paper (e.g. Kleenex). Flatten by gentle rubbing
    with a finger or glass rod. Replace the paper.
(3) Pour on a mixture of equal parts of chloroform, ethanol
    (100%) and xylene. Rub gently, replace paper and repeat.
(4) Repeat, using xylene.
(5) Mount, using plenty of mounting medium.

Methacrylate

Butyl methacrylate (p. viii) was the first embedding medium
successfully used for electron microscopy. It has since been largely
superseded for that purpose (see p. 61) but remains useful for
light microscopy, since it permits sectioning in the range 0.1 to
2 μm and can be used after fixation with glutaraldehyde or osmium
tetroxide. Excellent preservation can thus be combined with
sections of ideal thickness for phase-contrast or high-resolution
bright-field microscopy. For details of procedures see any of the
standard books on electron-microscope methods (e.g. Kay, 1965;
Pease, 1964; Sjöstrand, 1967).

Frozen sections

Frozen sections can be obtained rapidly, since little or no treatment of the tissue is necessary before sectioning. They are also
useful for histochemical work on lipids, enzymes and antigens,
which are damaged by normal fixation and embedding procedures.

35

SECTIONING METHODS

Less shrinkage occurs with frozen sections than with paraffin, and the disruption of the tissues caused by freezing and sectioning allows stains and reagents to penetrate readily into the cells. The disadvantages are that ice-crystal artifacts are usually present and serial sections and thin (5 μm) sections are difficult to obtain.

Preliminary treatment

The simplest procedure is to use fresh or formaldehyde-fixed tissues. Small (3 mm thick) pieces of such tissue are placed in a drop of water, saline or gum arabic on the chuck of the freezing microtome and attached by being frozen into place. Some tissues may section better if they are first embedded in 12 or 25% gelatin (Baker, 1944) or 2% agar. Tissues fixed in solutions containing mercuric chloride are washed first in a solution of iodine in ethanol (p. 22) and then taken to water. Those fixed in ethanol should be washed in water, otherwise they will not freeze.

Apparatus

Most modern freezing microtomes produce a low temperature in the specimen and knife edge by thermoelectric (Peltier) cooling. Microtomes that use carbon dioxide cooling are less satisfactory, since the cooling is less controllable and the specimen fluctuates between being too cold and too warm. This causes growth of large ice crystals in the specimen and enzymes may be inactivated. Dangerous explosions may be produced if the tube connecting the microtome to the CO_2 cylinder is too long.

The most satisfactory apparatus for producing frozen sections is the *cryostat*, which consists of an ordinary microtome in a refrigerated cabinet, with remote controls for the microtome. The advantages of the cryostat are that the temperature in the cabinet can be kept constantly at the optimum for sectioning, the specimen is not warmed by ambient air, and ribbons can be produced of material embedded in gelatin or agar.

Sectioning technique

The temperature of sectioning is usually between -15 and -20 °C and is critical. If the temperature is too low the sections may crumble, shatter or roll up excessively (cryostats are usually

36

equipped with a Teflon-coated guide plate that lies alongside the knife edge and prevents the sections from rolling up). If the temperature is too high the sections collect on the knife edge as a pulp.

When operating a CO_2 freezing microtome make sure that the CO_2 cylinder is vertical, with the valve at the bottom but above the level of the microtome. Short (2 s) bursts of liquid CO_2 are passed through the microtome.

A dry paint brush is used to transfer frozen sections from the knife edge to the fixative, stain or incubation mixture for further processing. Alternatively, a clean slide can be touched against the sections, so that they melt and adhere. No adhesive is necessary.

Non-frozen sections

Some fresh or fixed materials can be cut without being frozen or embedded. The sections ('non-frozen sections') have the advantages of frozen sections but morphological preservation is better, since ice-crystal artifacts are avoided. Such sections can be cut on two instruments: the Smith–Farquhar Tissue Chopper TC2 and the Oxford Vibratome. Smith (1970) compares the two instruments.

Decalcification

Bone and other calcified tissues are difficult to section unless first decalcified. The numerous variations of technique are discussed in detail by Drury & Wallington (1967) and Brain (1966). The two main methods are by the use of acids and chelating agents. Both are applied to fixed tissues, cut into thin pieces (e.g. slices of bone 3–5 mm thick).

Acids

5–10% nitric acid gives rapid decalcification (i.e. in about 4–24 h). Perenyi's fluid is less drastic and is generally useful.

Nitric acid (10%)	40 ml
Absolute ethanol	30 ml
Chromic acid (0.5%)	30 ml

37

In both cases transfer tissue to 70% ethanol after decalcifying and embed as usual.

Formic acid (8% in distilled water) decalcifies more slowly than nitric acid and causes less damage to tissues. Pieces of bone take 2–20 days, depending on size and density.

Chelating agents

EDTA (ethylene diamine tetra-acetic acid) combines with calcium to form a non-ionised substance. It works at neutral pH and is preferable to acids as a decalcifying agent, causing less damage and leaving tissues in better condition for staining. Its disadvantage is that it works slowly, decalcification taking from 4–40 days. Drury & Wallington (1967) recommend the following.

EDTA (sodium salt)	250 g
Distilled water	1750 ml

The solution is brought to pH 7 by addition of sodium hydroxide pellets. Use a volume of solution 20–30 times that of the specimen and renew every 5–7 days. Transfer to 70% ethanol when decalcification is complete and embed as usual.

Burstone (1962) discusses the effect of decalcification procedures on enzyme activity to be demonstrated by histochemical methods.

Staining methods

General notes

The methods described in this section are general staining techniques, designed to reveal structure. They do not usually provide any chemical information. For histochemical methods see the following chapter. For the theory of staining see Baker (1958). For other staining procedures see the standard handbooks (e.g. Conn, Darrow, & Emmel, 1960; Lillie, 1965; Drury & Wallington, 1967).

To avoid ambiguities in the naming of stains, the Colour Index numbers are given in Appendix 3. The number is taken from the 2nd edition of the list drawn up by the Biological Stain Commission. Conn (1961) gives the names, synonyms, Colour Index numbers and chemical formulae of the stains used in microscopy.

Stains whose coloured part is a cation (basic dyes) stain acidic, anionic substances, for example, nucleic acids. Such staining is often enhanced in alkaline solutions. Common basic dyes are methylene blue, toluidine blue, basic fuchsin, crystal violet, safranin, neutral red and pyronin. Stains whose coloured part is an anion (acid dyes) often stain all components of the tissue, especially at low pH. They are used, therefore, as counterstains. Examples are eosin, fast green, light green, aniline blue (water soluble) and ponceau 2R.

Before staining, sections are brought to water as described on p. 32.

Choice of method

As reliable general staining methods for animal material use Mayer's haemalum (p. 40), chlorazol black (p. 46), azure A/eosin B (p. 44) or Mann's methyl blue/eosin (p. 45). Mallory and Heidenhain's Azan method (p. 43) are also general histological staining methods and produce a particularly brilliant range of colours. Heidenhain's iron haematoxylin (p. 41) gives superb results at both the histological and cytological levels. It is good for nuclei, as is Hansen's trioxyhaematin (p. 42) and the special methods given in the section on nuclei and chromosomes. Sections of higher plants are best stained with safranin and fast green (p. 46) or chlorazol black (p. 46). Fungi are commonly stained with the lactophenol cotton blue method (p. 47). To show up the finest details of cells in the light microscope use Wigglesworth's osmic gallate technique (p. 48): the end result looks like a very low-power electron micrograph. Cytological details of plants and animals are also well shown in material embedded in plastic and stained by the method given on p. 65. Whole mounts of insects, small invertebrates etc. are stained with borax carmine (p. 47).

Mayer's haemalum

This method is quick and reliable. Do not use after osmium-containing fixatives.

A.

Haematoxylin	1 g
Sodium iodate	0.2 g
Potassium alum	50 g
Distilled water	1000 ml

Shake frequently until the solution is blue-violet. Then add:

| Chloral hydrate | 50 g |
| Citric acid | 1 g |

The solution now turns red–violet. Store in a hard glass bottle.

B. Eosin Y (saturated solution in 90% ethanol).

(1) Stain sections in haemalum, 5–15 min. The time is not critical as the stain is self-differentiating.

(2) Wash thoroughly in running tap water until the sections are blue (about 10 min).

(3) Take through 70% to 90% ethanol.

(4) Stain in eosin, 2–5 min.

(5) Wash in 90% ethanol to differentiate.

(6) Take through absolute ethanol, then xylene, 2 min each. Mount.

Nuclei are stained blue, cytoplasm pink.

As an alternative counterstain that shows up collagen use modified van Gieson's stain. Stain in haemalum and wash as above (steps (1) and (2)), then stain for 2 min in:

Ponceau S (1% aqueous)	100 ml
Picric acid (saturated aqueous solution)	900 ml
Acetic acid (2%)	15 ml

Rinse briefly in water, dehydrate, clear in xylene and mount. Collagen is stained pink, cytoplasm yellow.

Iron haematoxylin

This stain is especially good for material that is to be photographed. It is capable of resolving the finest details of structure in cells and tissues. Excellent after Flemming.

A. Iron alum (ferric ammonium sulphate) (3% in water).

B. Haematoxylin (5% in 96% ethanol)	1 vol
Distilled water	9 vols

The stock solution of haematoxylin should be allowed to 'ripen' (i.e. oxidise) for some weeks before use. Alternatively, add 0.2 g sodium iodate for each 1 g haematoxylin. Ripening is then effected in a few hours.

(1) Mordant sections in iron alum, 30 min to 24 h.
(2) Rinse in distilled water
(3) Stain in iron haematoxylin, 30 min to 24 h. It is usually recommended to stain for the same length of time as the sections were mordanted.
(4) Differentiate in iron alum. One may use the same (3%) solution that was used for mordanting, or dilute it to 1.5%. Differentiation must be controlled by removing the slide from the iron alum, transferring it to tap water (which stops differentiation) and examining under the microscope. Proceed until nuclear and cytoplasmic details are clear. This needs practice.
(5) Wash for one to several hours in running tap water.
(6) Dehydrate, clear and mount.

Nuclei, chromosomes and red blood cells are stained intensely black; other structures in shades of grey and blue–black.

Hansen's iron trioxyhaematin

This stains nuclei a clear black and leaves the cytoplasm unstained.

A.	Iron alum (ferric ammonium sulphate)	10 g
	$(NH_4)_2SO_4$	1.4 g
	Distilled water	150 ml

Heat gently to dissolve.

B.	Haematoxylin	1.6 g
	Distilled water	75 ml

Heat gently to dissolve.

Cool the solutions. Pour B into a large porcelain evaporating dish and add A to this, stirring constantly. Heat slowly, without stirring, just to boiling point. Cool rapidly by surrounding the dish with cold water. The originally deep violet solution should become dark brown, without any green sheen. Filter into small plastic containers that can be filled completely, to prevent oxidation, and store at −20 °C.

(1) Stain sections, 1–10 min.

(2) Wash in tap water, 15–30 min.

(3) As counterstain Masson's trichrome stain (see below) is suitable.

Masson's trichrome stain

This is a good counterstaining method after iron haematoxylin or Hansen's iron trioxyhaematin. Do not use after osmium fixation.

A.	Ponceau 2R *or* Ponceau S *or* Biebrich scarlet	0.25 g
	Acetic acid (1 % aqueous)	100 ml
B.	Phosphomolybdic acid	5 g
	Phosphotungstic acid	5 g
	Distilled water	to 100 ml
C.	Fast green (F.C.F.)	2.5 g
	Acetic acid (2 % aqueous)	100 ml

(1) Wash sections stained with iron haematoxylin or iron trioxyhaematin in running tap water, 15 min.

(2) Stain in A until rather darker than finally required (1–5 min).

(3) Rinse in distilled water.

(4) Treat with B, 1 min.

(5) Rinse in distilled water.

(6) Stain in C, about 2 min, examining in distilled water at intervals.

(7) Dehydrate rapidly, clear and mount.

Nuclei are stained black, collagen and mucus green, cytoplasm orange or pink.

Mallory

This stains tissues brilliantly red, blue and yellow, with some intermediate shades. It fades markedly within a year. Do not use after osmium-containing fixatives. The following method (Cason, 1950) is a rapid one.

Phosphotungstic acid	1 g
Orange G	2 g
Aniline blue (W.S.)	1 g
Acid fuchsin	3 g
Distilled water	200 ml

Add the ingredients to the water in the order given and dissolve each before adding the next.

(1) Stain for 5 min.

(2) Wash in running tap water, 3-5 s.

(3) Dehydrate rapidly in alcohols, clear in xylene and mount.

Nuclei red, nucleoli yellow, collagen blue, mucus blue, red blood cells yellow, cytoplasm pink or yellow.

Heidenhain's Azan

This gives results similar to Mallory but the staining is more precise and does not fade. Excellent after fixation in Susa; less good after Bouin, not recommended after osmium.

A. Azocarmine G *or* azocarmine B (0.1% in distilled water)

Boil. Filter when cold through soft filter paper. Much of the stain is in suspension and a hard filter paper may remove too much of it. Add 1 ml acetic acid (glacial) per 100 ml.

43

B. Aniline	1 ml
Ethanol (90%)	1000 ml
C. Acetic acid (glacial)	1 ml
Ethanol (96%)	100 ml
D. Phosphotungstic acid (5% in distilled water), freshly prepared	
E. Aniline blue	0.5 g
Orange G	2 g
Distilled water	100 ml
Acetic acid (glacial)	8 ml

Add the acetic acid last. Boil, filter when cold and dilute with twice the volume of distilled water.

(1) Stain in azocarmine for 1 h at 60 °C (30 min is long enough if the stain is already hot).

(2) Wash in distilled water.

(3) Differentiate in aniline alcohol (B) under the microscope until only the nuclei are stained.

(4) Wash in acetic alcohol (C).

(5) Treat with phosphotungstic acid (D), 1–3 h.

(6) Wash rapidly in distilled water.

(7) Stain in aniline blue–orange G (E), 1–3 h.

(8) Rinse very briefly in water or go straight to (9).

(9) Dehydrate in absolute ethanol, clear in xylene and mount.

Results as for Mallory (p. 43).

Azure A/eosin B

This method (Lillie, 1965) stains tissues in a large range of colours and is very reproducible. It is useful for pathological tissues as it shows up bacteria, parasites and necrosing tissues.

Azure A (0.1% aqueous)	4 ml
Eosin B (0.1% aqueous)	4 ml
Acetic acid (0.2 M)	1.7 ml
Sodium acetate (0.2 M)	0.3 ml
Acetone	5 ml
Distilled water	25 ml

Use the mixture once and then discard. The pH of this solution is 4.0 and is best for material fixed in formaldehyde. For material fixed in Zenker, use 1.25 ml acetic acid and 0.75 ml sodium acetate (pH 4.5); for material fixed in Bouin, Carnoy or Susa, use 0.7 ml acetic acid and 1.3 ml sodium acetate (pH 5.0–5.5). For solutions of higher pH see p. 110.

(1) Stain sections 1 h. (Celloidin sections need 2 h.)

(2) Dehydrate in acetone, 3 changes, clear in equal volumes of acetone and xylene, then in pure xylene.

(3) Mount in a synthetic resin (e.g. D.P.X.).

Since azure A is a basic stain, increasing the alkalinity of the mixture increases the blue staining.

Nuclei, RNA, bacteria stain blue; calcium deposits dark blue; mast cells and basophil granules, blue-violet; cytoplasm of most cells, pale blue; cytoplasm of necrosing cells and normal muscle cells, pink; cartilage matrix, red-violet; bone, pink; red blood cells, orange-red; mucins, green-blue to blue-violet.

Mann's methyl blue/eosin

This method stains tissues in the same colours as the azure A/ eosin B method, though all the stains used here are acid dyes. This is a good general method for sections of insects.

A. Methyl blue (1% aqueous)   35 ml
   Eosin (1% aqueous)   45 ml
   Distilled water   100 ml

Add a few drops of formalin as preservative.

B. Ethanol (70%) with one drop of a saturated aqueous solution of orange G per ml (Dobell's differentiator).

(1) Stain overnight.

(2) Rinse well in distilled water (20–30 s).

(3) Differentiate in B.

(4) Dehydrate rapidly, clear and mount.

For rapid work, stain 10–30 min, differentiate in tap water, dehydrate rapidly, clear and mount.

Chlorazol black

This metachromatic stain (p. 48) colours plant and animal tissues in shades of black, yellow, green and red. The solvent is usually 70% ethanol but can be water. The solvent markedly affects the colours that are developed.

Use a fresh, 1% solution of chlorazol black E in 70% ethanol. Do not filter.

(1) Bring sections to 70% ethanol.

(2) Stain 5–30 min.

(3) Dehydrate, clear and mount.

Plant cell walls stain jet black; nuclei may be black, yellow, or green; suberin stains amber; chitin greenish black; glycogen pink or red.

Safranin and fast green

This is for hand or paraffin sections of fixed plant material.

A. Safranin O (1% in 95% ethanol)

Dilute with an equal volume of distilled water before use.

B. Acetic acid (glacial)   1 ml
   Ethanol (70%)   100 ml

C. Fast green *or* light green
   (0.5% in clove oil/absolute ethanol (1:1))

D. Clove oil   50 ml
   Absolute ethanol   25 ml
   Xylene   25 ml

(1) Overstain sections in safranin, 1–24 h.

(2) Wash in distilled water.

(3) Differentiate in acid alcohol (B).

(4) Take through 90% ethanol to absolute ethanol.

(5) Stain in fast green, 30 s to 4 min.

(6) Differentiate in the clove oil, ethanol, xylene mixture, 2 changes of 5–15 min each.

(7) Pass sections through xylene and mount.

Nuclei, chromosomes, cuticle, and lignin stain red, other constituents green.

Lactophenol cotton blue

This method is commonly used to stain fungi, especially those invading plant tissues.

Phenol (pure crystals)	20 g
Lactic acid (sp. gr. 1.21)	20 g
Glycerol	40 g
Distilled water	20 ml
Cotton blue (= methyl blue)	0.05 g

Dissolve the phenol in water by warming gently, then add the lactic acid, glycerol and cotton blue.

Mount sections of plant tissues, or whole fungi, directly in this medium. Preparations are permanent.

Cotton blue stains the fungal protoplast blue. Lactophenol has a refractive index of 1.45, which makes fungal hyphae difficult to see. It also shrinks the stained protoplast considerably, so that fungal tissues appear abnormally small.

The cell walls of fungi can be stained with the PAS technique (p. 54).

Fungi growing on leaf surfaces can be stripped off with adhesive tape (Sellotape, Scotch tape) and stained with lactophenol cotton blue or PAS while still attached to the tape.

Borax carmine

Used for whole mounts. The stain is transparent and after differentiation usually stains mainly nuclei. After formaldehyde fixation, however, nuclei do not stain strongly.

Carmine	3 g
Borax	4 g
Distilled water	100 ml

Boil together for 30 min, cool and add an equal volume of 70% ethanol. Filter before use.

(1) Place the fixed specimen in the stain, 10 min.
(2) Differentiate thoroughly in acid alcohol (4 drops of concentrated HCl in 100 ml 70% ethanol).

(3) When the specimen is a bright transparent colour, dehydrate, clear and mount.

Osmic gallate

With this method (Wigglesworth, 1959) the finest cytoplasmic details are shown up in shades of grey and blue. As the staining is very intense, thin sections are necessary.

    A. Osmium tetroxide (1% in isotonic acetate-veronal buffer, pH 7.2 (p. 114))

    B. Ethyl gallate (progallin A from Nipa Laboratories Ltd, Treforest Industrial Estate, near Cardiff), saturated solution in 0.25% cresol (p. viii).

(1) Fix small pieces of tissue in osmium tetroxide (A), 4 h.

(2) Rinse in distilled water.

(3) Transfer to ethyl gallate solution, 16–24 h.

(4) Section. In Wigglesworth's original procedure, material is embedded in agar followed by ester wax (p. 32). Alternatively, embed in Araldite (p. 61) and section at 0.5–1 μm on an ultramicrotome, using a glass knife. Mount ester wax sections in Farrant's medium (p. 58) to which ethyl gallate has been added. Araldite sections can be mounted in immersion oil or resin medium.

Metachromasy

Some pure dyes will stain tissues in a range of colours if the tissue is examined in aqueous solution. This phenomenon is known as metachromasy. Both toluidine blue and methylene blue, for example, may stain some tissue components red, others purple and others blue. The substances that stain in the metachromatic colours (red and purple in the case of toluidine blue) are those that bind most dye. This is not obvious, however, in stained sections, since the metachromatic colours are paler than the colour of the free dye molecules. Metachromasy depends, in fact, on the binding of sufficient dye molecules for them to be close enough to interact with each other and with water (Bergeron & Singer, 1958). Thus metachromasy is influenced by all the factors that influence dye binding (e.g. dye concentration, pH, temperature, presence of competing salts etc.). The configuration of the stained substance, moreover, influences its ability to stack dye molecules. If these are stacked close enough for them to interact with each other, and water is present, metachromatic colours will be produced.

Metachromasy of basic dyes can be used to investigate the number or clustering of acidic groups on the substrate at a particular pH if all other variables are kept constant. The substrates most commonly investigated in this way are acidic polysaccharides and nucleic acids. It can also be used to investigate those properties of the substrate that allow the interaction of bound dye molecules (Feder & Wolf, 1965).

Acridine orange fluoresces metachromatically. This property is used for the same purposes as the metachromasy of other basic dyes (Stone & Bradley, 1961).

The following is a standard method for staining with toluidine blue.

(1) Bring sections to water.
(2) Stain in 0.5% toluidine blue, 4–6 h.
(3) Rinse in distilled water.
(4) Examine in water or mount in glycerine jelly.

Histochemistry

General note

For a comprehensive account of histochemical techniques see Pearse (1960, 1968). Here only some standard methods for the main classes of macromolecules are described.

Nucleic acids

DNA: the Feulgen technique

A. *Schiff's solution.* Dissolve 1 g basic fuchsin (p. viii) in 200 ml boiling distilled water. Stir well. Allow to cool to 50 °C and filter. Add 20 ml 1 M HCl to the filtrate. When cooled to about 25 °C add 1 g sodium (or potassium) metabisulphite ($Na_2S_2O_5$). Stand in the dark for 24 h, then add 2 g activated charcoal and stir for 1 min. Filter and keep the filtrate in the dark at 0–4 °C.

The solution keeps for 2 weeks and should be completely colourless. Allow to reach room temperature before use. Different batches of basic fuchsin differ in their intensity of staining when in Schiff's solution. A simpler and more rapidly prepared Schiff's solution is described by Kasten & Burton (1959).

See Humason— p. 332 (gives procedure)

B. *Bisulphite wash.* This should be freshly prepared.

Sodium (or potassium) metabisulphite ($Na_2S_2O_5$) (10%)	5 ml
1 M HCl	5 ml
Distilled water	90 ml

Any fixative except Bouin can be used. Formaldehyde is satisfactory.

(1) Bring sections or smears to water.
(2) Hydrolyse with 5 M HCl at room temperature. The optimum time will depend on the fixative that was used. Alcoholic fixatives will need 20 min–2 h; formaldehyde, 40 min–4 h; fixatives containing heavy metals, 2–4 h; freeze-dried material fixed in formaldehyde vapour, 2–8 h. The optimum time must be found by trial and error.

Hydrolysis can also be carried out with 1 M HCl at 60 °C for 5–40 min but the intensity of the colour developed is less.

(3) Rinse in distilled water.

(4) Transfer to Schiff's solution, 30–60 min.

(5) Wash in 3 changes of bisulphite wash.

(6) Rinse rapidly in water and counterstain if desired in fast green (p. 46).

(7) Dehydrate, clear and mount.

DNA stains red. The method is specific and stoichiometric. For controls omit hydrolysis or digest with DNase (see below).

RNA: methyl green/pyronin technique (Scott, 1967)

A. Dissolve 1 g methyl green in 100 ml 0.05 M sodium acetate buffer, pH 5.6 (p. 110). Shake with an equal volume of chloroform in a separating funnel. Repeat until the chloroform layer is colourless.

B. Dissolve 1 g pyronin Y (G) in 100 ml 0.05 M sodium acetate buffer, pH 5.6. Extract with chloroform as above.

Different batches of stain vary considerably in their purity and specificity of staining. Kasten (1967) found pyronin Y from Geigy gave consistently good results.

To prepare the staining solution mix 15 ml A with 25 ml B and make up to 100 ml with 0.05 M buffer, pH 5.6. Dissolve 40.6 g $MgCl_2.6H_2O$ in the final solution.

Fix in formaldehyde, Carnoy or 70% ethanol.

(1) Bring sections to water.

(2) Stain overnight (approximately 16 h).

(3) Rinse briefly in distilled water.

(4) Dehydrate in 2 changes of *n*-butanol, 5 min each.

(5) Clear in xylene and mount in D.P.X.

RNA stains red; DNA green.

For controls digest separate sections with RNase and DNase:

A. Dissolve crystalline deoxyribonuclease (0.05 mg ml^{-1}) in 0.1 M tris buffer, pH 5.7 (p. 113) containing 0.2 M $MgSO_4$.

B. Dissolve crystalline ribonuclease A (protease-free) (1 mg ml^{-1}) in phosphate buffer, pH 6.4 (p. 113).

Treat sections after step (1). Incubate for 24 h at 37 °C in DNase and 3 h at room temperature in RNase. After either treatment rinse in 5 changes of distilled water, then proceed as before.

An alternative method for RNA is to stain with azure B, with RNase digestion as control (Shea, 1970). Staining is said to be stoichiometric.

Proteins

There is no specific histochemical test for all proteins. In practice, specific tests for particular amino acids are used to reveal protein. Individual proteins are best demonstrated by immunofluorescent methods or, if the protein is an enzyme, by enzyme histochemical methods. For both techniques see Pearse (1968).

Tyrosin: the Millon technique (Baker, 1956)

Millon reagent

$HgSO_4$ 10 g
H_2SO_4 (10%)   100 ml

Heat until dissolved and make up to 200 ml with distilled water. Cool and add 20 ml of 0.25% $NaNO_2$.

Fix tissues in formaldehyde.

(1) Bring sections to water.
(2) Place sections in the reagent and boil gently.
(3) Cool to room temperature.
(4) Wash in 3 changes of distilled water.
(5) Dehydrate, clear and mount.

Proteins containing tyrosin are stained reddish. However, all phenolic compounds unsubstituted in the position *meta* to the hydroxyl group will react.

Arginine: the Sakaguchi reaction (Baker, 1947)

NaOH (1%)   2 ml
α-naphthol (1% in 70% ethanol)   2 drops
'Milton' (1%) (a proprietary brand of stabilised
sodium hypochlorite, containing approximately

1 g NaOCl and 18 g NaCl per 100 ml
(Pearse, 1968))   4 drops

(1) Bring sections to water.
(2) Allow sections nearly to dry out.
(3) Stain with α-naphthol/hypochlorite solution, 15 min.
(4) Drain and blot dry.
(5) Mount in pyridine plus chloroform (1 : 1)

The method is specific for arginine, which stains orange-red.

Lipids

Sudan black B (Chiffelle & Putt, 1951)

Most lipids can be stained with this.

Dissolve 0.7 g Sudan black B in 100 ml pure propylene glycol
by heating to 100–110 °C. Stir vigorously and do not let the
temperature rise above 110 °C. Filter hot through Whatman no. 2
filter paper. Cool and filter again at room temperature. (This takes
a long time: a fritted glass filter and a suction pump may be used.)

(1) Cut frozen sections of fresh or formaldehyde-fixed material.
(2) Wash sections in water, 2–5 min.
(3) Dehydrate in pure propylene glycol, 3–5 min. Because of the
    viscosity of propylene glycol, the sections (which are best
    stained loose) should be stirred in this and subsequent steps.
(4) Stain in Sudan black B, 5–7 min.
(5) Wash in distilled water, 3–5 min.
(6) Mount in glycerol jelly.

Lipids, unless masked (see below), are stained black.

Oil red O

This can be used for neutral lipids.

Dissolve 0.5 g oil red O in 100 ml 98% isopropanol. Before use
dilute 6 ml with 4 ml distilled water, allow to stand for 24 h and
filter through Whatman no. 42 filter paper immediately before use.

(1) Fix in formaldehyde and cut frozen sections.
(2) Rinse in freshly prepared 60% isopropanol.

(3) Stain in oil red O, 10 min.

(4) Rinse in water.

Neutral lipids are stained red. Nuclei can be stained blue with Mayer's haemalum (p. 40). Sections may be mounted in glycerine jelly.

For other methods for lipids see Pearse (1968). For a simple method of unmasking 'bound' lipids see Wigglesworth (1971).

Carbohydrates

Periodic acid/Schiff (PAS) technique

This will stain almost all materials containing sugars with un-substituted 1:2 glycol groups. Some of these materials (e.g. mono-saccharides) are lost during fixation.

    A. Periodic acid (HIO_4) (1 % aqueous)

    B. Schiff's solution (p. 50)

Any fixative is suitable, but if an aldehyde-containing fixative is used it is best to treat sections for 20–30 min at 22 °C with:

Aniline	10 ml
Acetic acid (glacial)	90 ml

If mercuric salts were present in the fixative wash sections with 96% ethanol made light brown with iodine.

    (1) Bring sections to water.

    (2) Oxidise in periodic acid, 10 min.

    (3) Wash in running water, 5 min.

    (4) Stain in Schiff's solution, 10 min.

    (5) Wash in running water, 5 min.

    (6) Nuclei can be counterstained with Mayer's haemalum (p. 40) but all other counterstaining should be omitted. Dehydrate, clear and mount.

For controls, omit the oxidation with periodic acid.

Chitin

There is no simple, specific histochemical test for chitin. Benjaminson (1969) describes a technique based on the binding of fluorescent-labelled chitinase; Richards (1951) and Pearse (1968) discuss available non-specific methods. For the chitosan test see p. 90.

Mucopolysaccharides

For methods see Pearse (1968). Acid mucopolysaccharides stain metachromatically (p. 48).

Cellulose

Staining with Schultze's solution is useful, though not specific.

Zinc chloride	50 g
Potassium iodide	16 g
Distilled water	17 ml

Dissolve, add an excess of iodine and allow to stand for several days. Decant the solution into brown dropping bottles.

Use fresh or fixed tissue. Do not use chromium-containing fixatives.

(1) Place sections in a few drops of the solution.

(2) Examine in the solution.

Walls with much cellulose stain blue; walls with much lignin, cutin, suberin or chitin stain yellow.

For an alternative, non-specific method for cellulose see under starch (below).

Starch

Dissolve 2 g potassium iodide in 100 ml distilled water, then dissolve 0.2 g iodine in this.

Place sections in a drop of this solution for a few min. Starch stains blue-black; newly formed starch may be reddish purple.

If sections are left in the solution for 15 min or more and then irrigated with a drop of 65% H_2SO_4 and examined immediately, cell walls containing cellulose will be seen to be stained blue, those containing lignin yellow. The results are not histochemically specific.

Callose

This can be stained in fresh sections with a 0.005% aqueous solution of resorcin blue applied for 15 min. The method is not strictly histochemical and different samples of the dye vary considerably.

Lignin

Stain sections of fresh or fixed material in a saturated solution of phloroglucin in 20% HCl. Lignin stains reddish violet.

Pectic substances

These can be stained fairly specifically by the hydroxylamine-ferric chloride method of Reeve (1959).

Glycogen

Stain sections with PAS (p. 54). Digest controls for 1 h at 37 °C in 1% malt diastase in phosphate-buffered saline, pH 7, or in saliva. After digestion wash thoroughly in running water before staining.

Glycogen stains mahogany brown in a saturated solution of iodine in absolute ethanol. See Pearse (1968) for Best's carmine stain for glycogen.

Mounting

This section deals with media for mounting stained sections or smears. Media for living material are described on p. 104. For detailed discussions see Baker (1960) or Lillie (1965).

Resinous media

For stained preparations the mounting medium should if possible have the same refractive index as the section (n = approx. 1.54). Canada balsam and various other natural or synthetic resins are most commonly used and are entirely suitable, provided the sections can be dehydrated after staining. If they cannot, aqueous mounting media must be used.

Canada balsam (n = 1.52) is usually used as a solution in xylene (approximately 60–65% w/v). After staining, sections are dehydrated in two changes of 95% ethanol, followed by two of 100% ethanol, then a mixture of equal parts of 100% ethanol and xylene, and finally two changes of xylene. The slide is then drained, a drop of Canada balsam placed over the sections and a coverslip gently lowered on top. Avoid air bubbles. Dry on a hotplate, or in a warm place. Complete hardening takes several months. Opaque areas result from inadequate dehydration; the coverslip should be soaked off in xylene and the preparation dehydrated again.

Various *synthetic resins*, based on polystyrene and other substances, are also used as mounting media and have largely replaced Canada balsam. They have the important advantage of being neutral, which minimises fading. They are sold under various trade names (e.g. D.P.X.) and for the most part are used in a similar manner to Canada balsam. Sections are mounted from xylene.

Euparal (n = 1.48) is a proprietary mixture containing eucalyptol, sandarac, camsal and paraldehyde (G.B.I. Laboratories Ltd, Heaton Mills, Heaton Street, Denton, Lancs.). It sets hard and causes little fading. Sections may be mounted direct from 95% ethanol, thus avoiding absolute ethanol and xylene.

Aqueous mounting media

These are less permanent than resinous media and their refractive index is usually lower. Among numerous mixtures (see Lillie, 1965) the following are useful.

Glycerol jelly ($n = 1.42$)

Gelatine	10 g
Glycerol	70 ml
Phenol	0.25 g
Distilled water	60 ml

Soak the gelatine in water for 2 h, then add the other constituents and heat on a water-bath, stirring continuously until the mixture is smooth. Store in the refrigerator. Apply melted and use a warm slide and coverslip. For delicate specimens it may be advantageous to transfer from water via 50% glycerol.

Farrant's medium ($n = 1.42$)

Gum arabic	40 g
Distilled water	40 ml
Glycerol	20 g

Dissolve the gum arabic in the water and add the glycerol. Add 0.1 ml cresol (p. viii) or other preservative.

Apathy's medium ($n = 1.52$)

Gum arabic	50 g
Sucrose	50 g
Distilled water	50 ml
Thymol	0.05 g

Dissolve by warming gently. This medium has a higher refractive index than other aqueous mounting media, and sets hard.

For permanence, preparations mounted in these media should be sealed (e.g. with 25% gelatin followed by gold size, or with 20% polystyrene in xylene) to prevent drying up.

For *temporary preparations* water or 50% glycerol ($n = 1.40$) may be used as mounting media.

Electron-microscope methods

General notes

Except for very small objects (e.g. viruses, bacterial flagella, proteins) material to be examined in the electron microscope has to be either sectioned or fragmented, since thick specimens are not penetrated by the electron beam. No satisfactory method exists for the examination of living material. Contrast in the image depends on scattering of electrons by the specimen. The amount of scattering is related to the energy of electrons in the beam and the density of the specimen. Elements of low atomic number are ineffective in scattering electrons and hence the intrinsic contrast of biological specimens is low. Salts of heavy metals (most commonly lead, tungsten and uranium) are used as stains, to increase contrast. With suitable specimens a resolving power of better than 0.3 nm has been achieved, but with most biological specimens a resolution of better than 1 nm is not to be expected. For an account of the electron microscope see Haine (1961) or Hall (1966); for preparative methods and the operation of the microscope see Pease (1964), Kay (1965), Sjöstrand (1967) or Ruthmann (1970). Juniper et al. (1970) deal specifically with plant material.

The above notes apply to the conventional electron microscope, in which the specimen is studied by transmitted electrons. In the *scanning electron microscope* the specimen is scanned by a narrow beam and the image formed by amplification of the electrons reflected by the surface or emitted secondarily as a result of excitation. Scanning microscopes give valuable information about the three-dimensional shape of the specimen and are coming into increasing use for biological work. For brief introductions see Nixon (1971) and Echlin (1971 a); for techniques see Echlin (1971 b).

Preparation of thin sections

Fixation

The most commonly used method is to fix in glutaraldehyde and to follow this by post-fixation in osmium tetroxide. For other procedures see the references cited above. For the mechanism of glutaraldehyde fixation see Millonig & Marinozzi (1968) and Hopwood, Allen & McCabe (1970). For the mechanism of osmic fixation see Riemersma (1970).

(1) Glutaraldehyde is supplied as a solution (usually 25%) in water and is diluted to 3% or less for use. The stock solution is not highly stable and should be stored in the refrigerator. Avoid inhaling the vapour. Glutaraldehyde is usually buffered at neutral or slightly alkaline pH with phosphate (p. 113) or cacodylate (p. 110) buffer. Sucrose should be added, especially for animal material, to raise the tonicity. (Bone & Denton (1971) recommend that the tonicity without glutaraldehyde should be about 60% that of the blood or body fluid.)

Fixative for plant tissues

Glutaraldehyde (25%)	12 ml
0.1 M phosphate buffer, pH 7.4	88 ml

Fixative for mammalian tissues

Glutaraldehyde (25%)	10 ml
0.1 M cacodylate buffer, pH 7.2	90 ml
Sucrose	6.8 g

Fix small pieces of tissue for 30 min to 4 h. Larger pieces require longer and for most purposes prolonged fixation in glutaraldehyde is not harmful. Glutaraldehyde penetrates well and quite large pieces of tissue can be fixed and subsequently dissected further or trimmed down to suitable size for further processing. It is usual, though not essential, to fix at 0–4 °C. After fixation, wash for several hours in several changes of cold buffer (0.1 M), to which sucrose may be added to bring the tonicity to that of the fixative (see Maser, Powell & Philpott (1967) for osmolality of buffers, fixative mixtures etc.). Material may be stored for weeks in buffer in the refrigerator.

(2) Post-fix in buffered osmium tetroxide:

Acetate–veronal stock solution (p. 114)	1 vol
0.1 M HCl	1 vol
2% osmium tetroxide	4 vols
Distilled water	2 vols
Sucrose	0.015 g/ml^{-1}

Adjust the pH to 7.4 with 0.1 M HCl or the buffer stock solution. Fix for 1–4 h, then wash briefly in buffered sucrose or distilled water.

Embedding

Epoxy-resins are most widely used. They are stable in the electron beam, polymerise with little volume change and are relatively easy to cut. The commercially available epoxy-resins are Araldite (Ciba) and Epon (Shell). The procedures for handling these are described by Luft (1961). For Araldite embedding proceed as follows.

(1) Dehydrate in an ethanol series (50%, 70%, 85%, 95%), 10–20 min each (for plant material start with 25% ethanol). Then take through 3 changes of absolute ethanol, 15 min each.

(2) Meanwhile, prepare an Araldite mixture as follows:

Araldite resin (CY 212)	27 ml
Hardener (HY 964)	23 ml
Benzyl dimethylamine	1 ml

Warm the Araldite resin and the hardener in a 45 or 60 °C oven before mixing and measure them with a warm graduated cylinder. The benzyl dimethylamine is an accelerator and a number of other substances can be substituted for it (see Luft, 1961). The components must be mixed very thoroughly; failure to do this is the commonest cause of difficulties in sectioning. After mixing, leave the mixture to stand in the oven for a short while, to allow bubbles to rise. The complete mixture as above cannot be kept for more than about a day (in the refrigerator), but a mixture of the resin and hardener can be kept for several weeks without polymerising. Great care should be taken not to get Araldite on the skin: it is possible to become sensitised to it, and it is possibly carcinogenic.

It is difficult to clean glassware thoroughly after it has been used for Araldite. It is simplest to use disposable beakers for making up mixtures, and to measure it in a measuring cylinder kept for the purpose and cleaned with acetone after use.

(3) Take specimens through 2 changes of propylene oxide, 15 min each. This evaporates very rapidly and care must be taken not to let the material dry out. Decant and fill the vials quickly. Avoid contact and do not inhale the vapour.

(4) Transfer to a mixture of propylene oxide and the Araldite mixture, 1:1, and leave for 1 h in this in a stoppered vial.

(5) After 1 h, remove the lid or stopper of the vial and allow the propylene oxide to evaporate. This should be done in a fume cupboard (hood). Evaporation takes several hours, at least, and it is simplest to leave material overnight at this stage.

(6) When the propylene oxide has evaporated, transfer the specimens to fresh Araldite in either gelatin capsules or flat trays. The latter (which may be plastic or metal planchettes, or small boats made of Al foil) are used for flat embedding, following which it is possible to check the orientation of the specimen and to cut sections in a known plane, after remounting.

(7) Polymerise the Araldite at 60 °C for 24 h or more.

Other embedding media are described by Glauert (1965 a) and Ruthmann (1970).

Sectioning

The preparation of glass knives and the operation of the various kinds of microtome are best learned from practical demonstration. Pease (1964) and Glauert & Phillips (1965) give useful descriptions. The following notes on sectioning may be helpful.

(1) Glass knives have a limited length of life. Only about 20 thin sections of Araldite can be cut from any given part of the knife edge. The knife will be blunted, as far as cutting thin sections is concerned, by cutting one thick (i.e. 1 μm) section. Avoid touching the edge with anything when picking up sections. Do not allow the knife edge to become dry once it has been wetted.

(2) Thin sections are cut most readily from small blocks, trimmed so that the sides are about 0.25 mm long. Larger blocks require more

care and must be cut more slowly. It is important to trim the block faces cleanly; ragged faces, with adhering shreds of embedding medium, easily get wetted by trough liquid, which makes sectioning difficult. The block should be in the form of a short, truncated pyramid. Long, thin blocks should be avoided.

(3) Before being picked up sections should be stretched by treatment with chloroform or xylene vapour. A paint brush, dipped in one of these, is held over the sections floating in the trough. Watch through the binocular microscope while doing this and avoid over-stretching.

(4) For most purposes sections should show silver to pale-gold interference colours after expansion.

(5) Difficulties in sectioning are dealt with at length by Glauert & Phillips (1965). Make sure that block and knife are rigidly clamped. If a cleanly trimmed block fails to section properly when cut at the right speed with a new knife set at the correct angle, the most likely reason is that the block is too soft.

(6) A single bristle from a paint brush, stuck to a toothpick or glass rod, can be used to manipulate sections floating in the trough before they are picked up.

(7) Pick up sections either from above, by touching them with a grid, or from below by submerging a grid and raising it under them, while steadying the sections with a bristle. It is difficult to position sections accurately and to avoid folds.

Grids and support films

Araldite or Epon sections can be picked up on bare grids. This is recommended for high-resolution work with thin sections, since a support film reduces contrast and may be a source of instability. 400-mesh grids should be used for this; the sections are subsequently stabilised with a layer of evaporated carbon. For most purposes, however, it is preferable to use a celloidin or Formvar film stabilised with carbon. With such a support film grids with larger holes can be used (e.g. 200 or hexagonal mesh).

A simple method of preparing celloidin films is as follows. Grids are placed on a piece of stainless steel mesh submerged in a dish of distilled water. The dish should be about 15 cm in diameter and the mesh should be supported so as to lie 2–3 cm below the surface of the water. One or two drops of a 2% celloidin solution in dry amyl acetate (filtered) are dropped on the surface of the water. The solution spreads out and the solvent evaporates. Wait for a minute or two, then siphon water out of the dish until the film (which should be grey or pale silver when viewed by reflected light) makes contact with the grids. As an alternative to siphoning, the dish may be fitted with a tap at the bottom and the water allowed to run out. Lift the grids out on the mesh, drain and dry in a desiccator. When dry remove the grids from the mesh and coat with a thin layer of carbon in a vacuum coating unit. It is important to use a high-viscosity nitrocellulose for the celloidin solution.

Staining

Two commonly used stains are uranyl acetate and basic lead citrate. These may be used alone, or successively (double staining).

Uranyl acetate is dissolved in 50% ethanol to form a saturated solution. This should be kept in the dark and discarded if it develops an orange-red colour. Filter or centrifuge before use. Stain sections for 30–60 min by floating grids singly face down on drops of the stain on a piece of dental wax in a covered Petri dish. Keep sections out of direct sunlight while staining. Then wash thoroughly with distilled water, either by dipping or with a gentle stream from a wash-bottle.

Basic lead citrate (Reynolds, 1963) is one of numerous lead-containing stains. It is made up as follows. Dissolve 1.33 g lead nitrate and 1.76 g sodium citrate ($2H_2O$) separately, each in about 15 ml distilled water. Combine in a 50 ml volumetric flask, shake vigorously for 1 min and then agitate gently on a flask shaker for about 30 min. Add 8 ml carbonate-free 1M NaOH, agitating gently while doing so. The final solution should be clear. If necessary the solution may be centrifuged before use to remove turbidity. The solution keeps for several weeks. Stain grids on drops of the stain on a piece of dental wax. It may be advantageous to remove CO_2 from the atmosphere while staining by placing a pellet of sodium hydroxide in the Petri dish. Stain for 5–15 min. Then wash very carefully, first with 0.02M NaOH and then with distilled water.

Difficulties with staining arise chiefly from using old solutions or from inadequate washing. When washing sections make sure that stain is not trapped between the points of the forceps. Sections should preferably be stained fairly soon after cutting.

Thick sections for light microscopy

Sections (called 'thick' sections in this context) can be cut on the ultramicrotome at o.5–1 μm, for examination in the light microscope. This may be useful as a preliminary to thin sectioning, as a means of ascertaining rapidly the quality of fixation, of checking that the specimen is correctly orientated, or that the desired region is being cut. It also permits cytochemical observations to be made on sections adjacent to the thin sections to be studied in the electron microscope.

Thick sections should be cut slowly. They are picked up individually with a fine glass needle or a toothpick, broken so that the tip is formed by a thin splinter. Transfer each section to a very small drop of water on a slide and dry on a hotplate. No adhesive is necessary. Such sections may be rapidly stained by covering with a drop of 1 % methylene blue in 1 % borax and heating until the stain begins to steam. Wash off the excess stain with a stream of water or 50% ethanol and dry. The sections may be mounted in immersion oil and examined without a coverslip. Ruthmann (1970) describes other staining methods and Wanson (1964) gives a summary of applicable techniques.

Electron-microscope cytochemistry

Cytochemical techniques for nucleic acids, glycogen and mucopolysaccharides are described in Ruthmann (1970). Plant cytochemistry is dealt with by Juniper et al. (1970). Highly specific methods are available for the localisation of antigenic substances involving the use of antibodies labelled with either ferritin (Glauert, 1965 b) or peroxidase (see Kawarai & Nakane, 1970). Some enzymes can be stained. For electron-microscope autoradiography see p. 74.

Freeze etching

In this technique, rapidly frozen material is cut while still frozen and the outermost layer of ice is sublimed away in a vacuum from the cut surface of the block, thereby exposing cell constituents. This is termed 'etching'. A layer of carbon or other material is evaporated onto the etched surface, which is shadowed (see below). The shadowed replica is then floated off and examined in the electron microscope. The method yields valuable information about the configuration of membranes and other cell components and has the advantage of avoiding chemical fixation. A special apparatus is necessary. For discussion of techniques see Moor (1964, 1971).

Negative staining

This technique is used for the examination of proteins, viruses, and parts of cells such as microtubules, chromosomes or membrane fragments. The stain, which is opaque to electrons, is used to surround the specimen, which appears light against a dark background. Substances used as negative stains dry out without detectable structure. They may follow the contours of the specimen very closely and penetrate into the smallest grooves etc., so that, for example, individual protein subunits of a virus particle are made visible.

The two most commonly used negative stains are potassium (or sodium) phosphotungstate and uranyl acetate. The former is made by adding 1M KOH to a 1% solution of phosphotungstic acid until the pH is 6.8–7.4. The latter is used as a 1% solution in distilled water. The stain may be mixed with a suspension of the material to be examined, sprayed onto grids with a nebuliser and allowed to dry. Alternatively, drops of the suspension and stain are placed on the grid with a pipette, the material is allowed to settle for a minute or two and then the bulk of the liquid is drawn off with filter paper. It is also sometimes effective to place the material on the grid first, let it dry down, and then cover it with a drop of the stain, which is drawn off with filter paper after about one minute. In all cases rapid drying is advantageous. Other negative stains are discussed by Horne (1965).

Some cell constituents (e.g. chromosomes, microtubules) can be prepared by the surface spreading method and then negatively stained (see Gall, 1966a). The cells are introduced onto a clean water surface, where they lyse. Some components remain on the surface and can be picked up on grids and stained as above.

Shadowing

This method is used for bringing out detail in surface replicas (for example in freeze etching; see above), or for examining small particles such as viruses or nucleic acid or protein molecules. It consists of evaporating metal from a source placed at an angle to the specimen. Metal is deposited generally over the grid, except in the shadow of the specimen, which therefore appears in the electron microscope outlined on one side by a light margin. If the angle of shadowing is accurately known the method can give valuable information about molecular shape and dimensions. For details see Bradley (1965).

Quantitative techniques

Sampling methods for the estimation of areas and volumes etc. are described by Loud (1962), Weibel & Elias (1967), and Weibel (1969).

Autoradiography

Principle

This technique can be regarded as a special type of histochemical procedure. It permits the localisation of particular substances by virtue of the fact that they specifically incorporate certain precursors, which are supplied labelled with a radioactive isotope. The location of the labelled substance so formed is made visible by covering sections, squashes etc. of the cells or tissues with a thin layer of photographic emulsion, which acts as a radiation detector. After exposure the emulsion is developed and fixed, like a photographic plate or film, and studied in the microscope. The section or squash remains in contact with the emulsion and the silver grains, marking sites of radioactive decay, can be related to underlying structures. The method is used for studies of the synthesis, transport, localisation and turnover of any substance which is retained in tissue sections and for which a suitably labelled precursor is available. It can be used at both the light- and electron-microscope levels. For full descriptions see Rogers (1967) or Schultze (1969).

Information on isotopes and radioactive compounds is contained in *The Radiochemical Manual* (1966, 2nd ed.). The β-particle emitted by tritium (^{3}H) is of low energy and is correspondingly able to penetrate tissue and emulsion for only a short distance. As a result, silver grains produced in the emulsion are located close to the site of decay, so that the autoradiographs give accurate information about localisation. This is one reason why tritium is the most commonly used isotope in autoradiography. It has the added advantages that it can be used to label a large number of biologically important substances, and that tritiated precursors of high specific activity[1] can be prepared. [^{3}H]thymidine is commonly used as a precursor for DNA, [^{3}H]uridine for RNA, and various tritiated

---

[1] The unit of radioactivity is the curie (Ci). 1 curie $= 3.7 \times 10^{10}$ disintegrations per second; 1 microcurie (μCi) $= 3.2 \times 10^9$ disintegrations per day. At a specific activity of 1 mCi m^{-1} 1 disintegration per day is given by 1.88×10^8 molecules.

amino acids for proteins. ^{14}C-labelled compounds are also commonly used. It cannot be taken for granted that the incorporation of such precursors is completely specific. Exceptions have been reported and it may be necessary to check specificity by digestion with enzymes, as with histochemical staining methods.

The basic procedures are simple (see Prescott, 1964). Refinements are necessary when high-resolution autoradiography is attempted (Caro, 1964) or when quantitative data are sought (Perry, 1964). In general, the resolution of autoradiographs is improved by working with thin specimens and thin emulsion layers, and by using an emulsion of small grain size. The best resolution so far obtained, even with electron-microscope autoradiography, is about 0.1 μm.

Procedures

Application of isotope

All radioactive isotopes must be handled with care. Local regulations concerning their use and disposal should be strictly observed. Direct contact or intake should be avoided; even weak sources, if incorporated in the body, are potentially dangerous.

Labelled compounds are usually supplied in sterile solution in rubber-sealed ampoules or bottles. The concentration and specific activity are usually stated. A calibrated syringe is used to withdraw the required amount. Isotopes are commonly diluted with saline (p. 104) or culture medium before being administered to experimental organisms. The method of administering them depends on the type of organism. For small aquatic organisms, algae, protozoa, bacteria or tissue-culture cells the isotope is simply added to the culture medium. Plants may be grown with their roots in medium containing labelled compound. The level of dosage required has to be determined by trial. The duration of exposure to labelled compounds depends on the object of the experiment. Sites of synthesis may be demonstrated by labelling for only a few minutes. Study of chromosome replication, in which it may be necessary to follow one or more rounds of chromosome division, require exposures of some hours. Transport of macromolecules may be studied by using brief exposure to labelled compound, lasting a few minutes (*pulse labelling*), followed by washing in an excess of unlabelled precursor (*chase*) to dilute the labelled precursor. The label is observed first at the site of synthesis and subsequently at the site (or sites) to which it is transported.

Note that labelled molecules sometimes fail to penetrate into cells or organisms and in some cases are not metabolised normally as precursors.

Tissue preparation

After exposure to isotope, cells and tissues are usually fixed. Any fixative can be used which gives adequate preservation of structure, but substances which may react with the emulsion (e.g. picric acid, formaldehyde) should be washed out thoroughly, or avoided. Do not use glutaraldehyde when studying incorporation of amino acids into proteins. Ethanol-acetic acid (p. 76) or Carnoy's fluid (p. 76) are suitable for light-microscope work; for electron-microscope autoradiography glutaraldehyde is used in the ordinary way (p. 60). Freeze-drying or freeze-substitution may be used for water-soluble substances, or cryostat sections may be used (see Miller, Stone & Prescott, 1964; Roth & Stumpf, 1969). Embedding and sectioning are done by standard procedures. For ordinary work paraffin is used and sections are cut at about 5 μm. Methacrylate or epoxy-resin may be used for preparing thinner sections (0.4 μm) for high-resolution work. For chromosomes use squashes (p. 77).

Slides should be cleaned very thoroughly. To ensure good adhesion of tissue and emulsion to the slides they may be coated ('subbed') with chrome alum gelatin (this step is not essential). Dissolve 1 g gelatin in one litre of warm distilled water; add 0.1 g chromium potassium sulphate when cool. Dip slides in this and then stand upright in a desiccator to dry. Sections are mounted in the usual way (though without Mayer's albumen). Paraffin must be completely removed with two changes of xylene.

Precursor not incorporated into macromolecules is largely removed during fixation. Any residue may be extracted by treating sections with 5% trichloracetic acid at 4 °C for 5 min, followed by thorough washing in 70% ethanol (3 changes, several hours in all).

Application of emulsion.

Emulsions may be applied either as a liquid or by the stripping film technique. Both methods have advantages and are widely used.

Stripping film technique. For this, the only commonly used emulsion is Kodak AR-10. This is supplied as a layer carried on gelatin, mounted on glass plates (figure 2*A*). It should be stored in the refrigerator and may be kept for up to 6 months. It can be handled with a safelight (Wratten no. 2). For coating slides the emulsion must first be stripped from the glass. Using a sharp knife, score

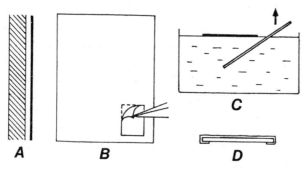

Figure 2. The stripping film technique. *A*, the film on its glass support; *B*, the method of scoring and peeling off the piece of film from the glass; *C*, the film floating emulsion-side down on water, with the slide submerged below it; *D*, the autoradiograph in section.

around a rectangular area large enough to cover the sections and to overlap three sides of the slide. Cut the first piece at one corner of the plate (figure 2*B*), leaving a margin of about 1 cm. Then, holding the plate vertically, peel off the piece of film slowly and steadily with forceps. Place it emulsion side down on the surface of a dish of water and allow it to absorb water for 3 min. Immerse the slide to be coated in the water, position it under the piece of film and lift it out, without trapped air bubbles (figure 2*C*). Stand vertically to drain and dry for 20–30 min, if possible in a stream of dry, cool air.

The stripping film technique has the advantages of simplicity and of yielding an emulsion layer of very uniform thickness. The latter makes it suitable for quantitative work.

Liquid emulsions. The ones in most general use are Ilford K5 and Kodak NTB-2 and NTB-3. For the characteristics of these and others see Rogers (1967) or the manufacturers' information. The emulsions are supplied as gels which melt at 42–45 °C. They can be handled in a red safelight. Emulsions can be stored for 2 months (preferably in the refrigerator) without significant increase in background. To coat slides with Kodak NTB-2 or NTB-3 the bottle of emulsion is first warmed at 43 °C in a water bath for 30 min and a suitable amount transferred to a beaker or cut-down measuring cylinder, also warmed in the water bath. The slides (preferably dry and at the same temperature as the emulsion) are then dipped for 4–5 s. They are withdrawn slowly and steadily, drained, the backs are wiped with tissue and the emulsion allowed to dry for a few hours. The aim is to produce a thin, uniform layer.

The procedure with Ilford emulsions is slightly different. The emulsion is taken from the bottle as a solid, using plastic print forceps, and is melted at 43 °C in a 50 ml measuring cylinder. One volume is then added to an equal volume of pre-warmed 2% glycerol, the mixture is stirred gently and allowed to stand for a few minutes to allow air bubbles to rise, and then slides are coated as described above.

Emulsion which has once been warmed is usually discarded. A test for background can be made by coating a blank slide and developing it as soon as it has dried.

The advantages of the liquid emulsion technique are that very thin films can be produced, the emulsion is in close contact with the specimen, and a range of emulsions is available for different purposes. The films adhere well and the gelatin that remains after processing is much thinner than with the stripping film technique. The chief drawback is that it is extremely difficult to produce films of constant thickness.

Exposure

Coated slides, when dry, are placed in light-tight boxes for exposure. Some workers place silica gel or other drying agent in the boxes and keep them at 4 °C during exposure, but this is not essential and excessive drying may be harmful. The optimum length of exposure depends on a number of factors and must be determined by trial. A slide can be developed from time to time to check if exposure has proceeded long enough. An exposure time of 1–3 weeks should be aimed at and the isotope concentration used should if possible be adjusted to achieve this.

Development

Follow the instructions given by the manufacturers of the emulsion. A fine-grain developer is essential. Kodak D19b is commonly used, for both stripping film and liquid emulsions:

4-methylamino-phenolsulphate ('Metol', 'Elon')	2.2 g
Sodium sulphite (anhydrous)	72.0 g
Hydroquinone (crystalline)	8.8 g
Sodium carbonate (anhydrous)	48.0 g
Potassium bromide	4.0 g

Dissolve each component completely, in the order given, in about 750 ml distilled water at 35–40 °C and make up to one litre.

Develop autoradiographs for 2–5 min at 18 °C. Rinse for 30–60 s in distilled water, fix (for example, in 30% sodium thiosulphate for 8–10 min), wash for 5–10 min in running tap water, rinse in distilled water and dry in air.

Microscopy

Autoradiographs can be mounted in glycerol and examined by phase-contrast without any further preparation. More usually they are stained. This is done either before coating, or through the emulsion after developing, with such stains as toluidine blue (p. 49), azure B or methyl green/pyronin (p. 51). A simple procedure is to stain for a few minutes in 0.25% toluidine blue at about pH 6, and wash thoroughly in water or 95% ethanol to remove excess stain. Dry in air and mount in Euparal.

A × 63 oil-immersion objective may be preferable to a higher-power lens for examining autoradiographs: its greater depth of focus permits both emulsion and underlying specimen to be brought simultaneously into focus.

Electron-microscope autoradiography

The principles are the same as for light-microscope autoradiography. Liquid emulsions are invariably used. The limit of resolution (about 0.1 μm) is set by the grain size of the emulsion. For descriptions of techniques see Caro (1964), Salpeter (1966), Stevens (1966) or Rogers (1967). There are many minor variations in technique.

Nuclei and chromosomes

Nuclei

The chromatin of nuclei can be stained with basic stains (p. 39). Among standard staining methods, iron haematoxylin (p. 41) and the gallocyanin technique (p. 96) are good for nuclei. Chromatin and nucleoli can be stained differentially by the methyl green/pyronin technique (p. 51).

Nuclei generally fluoresce green, nucleoli and cytoplasmic RNA orange when stained with acridine orange (Kasten, 1967). As a simple procedure use fresh smears or frozen sections and stain for 15 min with 1 in 10000 acridine orange, adjusted to pH 6.0–6.5 by addition of 0.1 M phosphate buffer. It may be necessary to experiment with solutions of different pH. Examine in water using the fluorescence microscope (p. 5).

Chromosomes

Preliminary treatment

Chromosomes are usually studied at metaphase, when they are at their shortest and thickest. Dividing cells can be obtained by growing cells in culture (p. 14), in plants by treating with hormones (such as indole acetic acid, 20–50 p.p.m. applied to roots for approximately 4 hours) and in mammalian small lymphocytes by treating with the plant extract, phytohaemagglutinin (Heuser & Razavi, 1970).

Separating the chromosomes from each other makes it easier to identify and count them. This may be achieved in four ways, which may be used separately or in combination.

(1) Treat with chemicals that prevent the formation of the spindle.

(a) *Colchicine or colcemid* (*n*-deacetyl *N*-methyl colchicine). Add an aqueous solution to the medium bathing the cells. The final concentration should be $10 \mu g \, ml^{-1}$ (0.001 % or $2.5 \times 10^{-5} M$). Treat for at least 3 h. It is effective on animal cells even at a final concentration of $0.5 \mu g \, ml^{-1}$ (Mueller,

Gaulden & Drane, 1971). For plant cells, concentrations as high as 0.05–0.2% are commonly used, but may cause supercontracted chromosomes.

(b) *Vinblastine sulphate* (Eli Lilly & Co., Indianapolis). Use at a final concentration of 0.1–1 μg ml^{-1} for 8–15 h (Maio & Schildkraut, 1966).

These agents arrest cells at metaphase; the longer they act the more metaphases are produced. After treatment for 24–48 h however, abnormalities of mitosis occur. For other chemicals used to produce metaphase arrest and chromosomal spreading see Sharma & Sharma (1956).

(2) Treat cells with a hypotonic solution to make them swell. For animal cells, dilute the medium surrounding the cells with double the volume of distilled water. Treat for 20 min. Alternatively, flood mammalian cells with 0.0975 M KCl solution for 5 min, or 1% sodium citrate containing 10^{-3}M MgCl$_2$ and 10^{-3} CaCl$_2$ for 30 min. Soft plant tissues can be treated with distilled water for a few min.

(3) Expose to low temperature. This interferes with the stability of spindle fibres. If plant tissues are kept at 0 °C for up to 2 days, the number of cells at metaphase is often increased and separation of the chromosomes is facilitated.

(4) Squash cells between slide and coverslip after fixation. This is described in the technique for staining with aceto-carmine (see below). Plant tissues may require maceration in equal parts of concentrated HCl and 95% ethanol for 5 min before squashing.

Fixation

Acetic-ethanol. Mix immediately before use 3 parts absolute ethanol and 1 part glacial acetic acid. Keep chilled, to avoid the formation of esters. Fix for 2 min to 24 h. Chromosomes are well preserved but cytoplasmic organelles are destroyed.

Carnoy. This is a similar fixative, prepared and used in the same way.

Absolute ethanol	60 ml
Chloroform	30 ml
Acetic acid (glacial)	10 ml

Formaldehyde. Use this for general fixation of cytoplasm as well as of chromosomes. Formaldehyde can be used before Feulgen.

Staining

Aceto-carmine

Acetic acid (glacial)	45 ml
Distilled water	55 ml
Carmine	1 g

Boil gently in a reflux condenser, 5 min. Shake well and filter when cool.

(1) Tease fresh tissue apart in a drop of the stain on a slide for a few min. If iron needles are used a darker brown-red stain is produced. Cover with a coverslip and warm the slide gently over a small flame.

(2) Place several thicknesses of filter paper over the coverslip and apply firm pressure with the thumb. Allow no sideways movement of the coverslip. Examine.

(3) Preparations can be made permanent. Place the slide, coverslip downwards, in a dish of 10% acetic acid. The coverslip will fall off in 15 min or so. Dehydrate the slide and coverslip separately in 1:3 acetic–ethanol, 2 min, then in absolute ethanol, 2 min. Take through xylene and mount.

In some plant material the cytoplasm may stain as well as the chromosomes. If this happens, fix the material for 12–24 hours in acetic-ethanol before staining. Alternatively, use the aceto-orcein stain.

Aceto-orcein. This is an excellent and rapid stain for almost all types of chromosomes. The quantity of orcein to use in making up the stain depends on the particular batch of dye and is best determined by trial and error. Try 1% aceto-orcein in 45% acetic acid, made up and used as aceto-carmine but refluxed for 30–60 min. The solution should be a deep purple colour, rather than red. A stronger solution will be needed of American synthetic orcein. Keep the dye solution in the refrigerator and centrifuge if it appears cloudy. The dye precipitates if the concentration of acetic acid is significantly reduced, as for example by a drop of water on the slide.

Feulgen. See p. 50. This may be used on sections or squashes of fixed material. Only DNA is stained.

Quinacrine. Individual chromosomes can be identified at metaphase by the pattern of fluorescent bands produced by quinacrine staining.

Dissolve 0.5 g quinacrine (atebrine or mepacrine) in 100 ml McIlvaine's citrate/phosphate buffer, pH 4.5 (p. 111).

(1) Fix tissues for a few min in acetic-ethanol. Smears can be fixed in methanol.
(2) Wash in distilled water for a few min.
(3) Stain for 5 min.
(4) Wash in running tap water, 3 min.
(5) Mount in buffer at pH 5.5, squash if necessary and examine with the fluorescence microscope.

Similar but more clear-cut staining can be achieved with quinacrine mustard (Polani & Mutton, 1971; Loveless, 1970).

Sex chromatin

The inactivated X-chromosome (Barr body) of females of certain species, including man, can be stained with aceto-orcein. Fixed material should be treated as described in Eggen (1965). The Y-chromosome of males of certain species, including man, can be stained in the resting nucleus or at metaphase, or in sperm, by the quinacrine or quinacrine mustard technique. See Sumner, Robinson & Evans (1971).

Methods for special cells

For *mammalian embryos* see Ford & Woollam (1963). For puffs of dipteran *salivary gland chromosomes* see Ashburner (1967). For *lampbrush chromosomes* of amphibia use phase-contrast microscopy and freshly isolated chromosomes (Gall, 1966b). For *yeasts* see Robinow (1970). For *blood cells* cultured to reveal chromosomal abnormalities see Heuser & Razavi (1970). Methods for *human chromosomes* are dealt with in Yunis (1965).

For *cultured animal cells*, after hypotonic treatment (p. 76) fix in acetic-ethanol (p. 76) and allow a drop of the suspension to fall from a height of several feet on to a cold, wet slide. Dry in air or by gentle warming. Stain with aceto-orcein (p. 77). Do not squash.

Mitotic spindles

These may be stained by many of the standard procedures; Weigert's haematoxylin (Drury & Wallington, 1967) is particularly useful. In living material spindles are most readily seen with the polarizing microscope (see Inoué, 1964).

Bacteria

General

For methods in bacteriology see Cruickshank (1969) and Norris & Ribbons (1969, 1970). For the theory and practice of culture methods, sterilisation etc. see Meynell & Meynell (1970). Cultures of bacteria can be obtained from the National Collection of Type Cultures, Central Public Health Laboratory, Colindale Avenue, London NW 6.

The methods described here are some standard techniques for the light-microscopic examination of bacteria. For electron microscopy see Kay (1965).

Nigrosin method

Useful for demonstrating bacterial size and shape, and often used for preliminary examinations.

Dissolve nigrosin in warm distilled water to make a 10 % solution. Filter and add 0.5 % formalin as preservative. Mix a small drop of the stain with the bacteria and make a thin smear. Allow to dry and examine using oil-immersion (no coverslip is necessary). The nigrosin forms an opaque background against which the bacteria stand out as light objects.

Capsules

These can be demonstrated most simply by mixing a small quantity of the bacterial culture with a drop of India ink on a slide. Add a coverslip and apply pressure through several thicknesses of filter paper. The ink film should spread out and look pale. Capsules appear as clear spaces between the bacterial cells and the dark background.

Gram's method

This is used both diagnostically and to demonstrate general morphology. In outline, bacteria are stained with crystal violet, followed by iodine, and then treated with acetone or other de-colorising agent. Gram-positive organisms retain the stain, Gram-negative ones lose it. The mechanism of staining is not fully understood but seemingly depends on both the chemical composition of the cytoplasm and the permeability of the cell wall. Gram-negative organisms are demonstrated by staining with basic fuchsin after Gram staining.

A. Crystal violet stain

a Crystal violet (1 % in distilled water)

b Sodium bicarbonate (5 % in distilled water)

Mix 30 vols of *a* with 8 of *b* before use.

B. Iodine   20 g

    1 M NaOH   100 ml

    Distilled water   900 ml

Dissolve the iodine in the NaOH solution and then add the distilled water.

C. Basic fuchsin (0.1 % in distilled water)

(1) Make smear and allow to dry, or dry in warm air. Fix by flaming (i.e. pass slowly 2 or 3 times through a Bunsen flame).

(2) Cover with crystal violet and stain for 5 min.

(3) Drain, and wash off excess stain with iodine poured over the slide. Then cover with fresh iodine solution and leave 2 min.

(4) Drain, and flood with acetone. Decolorisation takes only 2–3 s and the acetone must be washed off immediately by placing the slide under a running tap.

(5) Stain with basic fuchsin, 30 s.

(6) Rinse in tap water, 5 s, blot, and dry in air.

Gram-positive organisms appear dark violet, Gram-negative pink.

Chromatin bodies

(1) Fix smears on coverslips in osmium tetroxide vapour, 2–3 min. Alternatively, fix bacteria growing on pieces of agar in the same way and then place the agar face down on a coverslip. Dry in air.

(2) Fix in warm Schaudinn (p. 85), 5 min.

(3) Wash in water.

(4) Hydrolyse in 1 M HCl at 60 °C for 10 min.

(5) Rinse in tap water, then distilled water.

(6) Float face down on Giemsa stain, made by diluting the stock solution as purchased (e.g. G. T. Gurr's R66) with 0.005 M phosphate buffer (p. 113), pH 7.0, at the rate of 2–3 drops of stain per ml buffer. Stain 30 min at 37 °C.

(7) Mount in water and examine. Preparations may be sealed and retain their stain for a few days.

This method reveals chromatin bodies reliably. Instead of Giemsa, the Feulgen method for DNA (p. 50) may be applied after hydrolysis.

Protozoa

Introduction

Protozoa are single-celled organisms and the methods of studying their structure are not much different from those used for cells of multicellular organisms. Most standard cytological and cyto-chemical procedures can be applied with little or no modification. For comprehensive accounts of materials and methods see Kirby (1950) and Mackinnon & Hawes (1961).

Sources of material

A list of protozoa in culture was published in *J. Protozool.* **5**, 1–38 (1958). The principal culture collections of protozoa are as follows:

Algensammlung, Pflanzenphysiologisches Institut, Göttingen, West Germany.

Culture Centre of Algae and Protozoa, 36 Storey's Way, Cambridge CB3 oDT.

Culture Collection of Algae, Botany Department, Indiana University, Bloomington, Indiana, U.S.A.

Kirby (1950) and Mackinnon & Hawes (1961) both contain much information about obtaining and culturing protozoa.

Living material

The more delicate features of protozoa are often not preserved in fixed material and the study of living organisms is particularly important. Protozoa may move abnormally when restricted between slide and coverslip: a high-power stereoscopic microscope may be useful for studying freely swimming organisms in Petri dishes etc. For detailed studies of cytology, phase-contrast (p. 3) is in-valuable. As far as possible, protozoa should be studied in the medium in which they occur naturally or have been cultured. Physiological salines (p. 104) can be used for diluting gut fluids etc. when studying parasites. Wet preparations should be sealed with Vaseline or paraffin wax, applied around the edges of the coverslip with a warm needle or paint brush, to prevent evaporation.

Rapidly moving organisms may be immobilised by progressively withdrawing fluid, using filter paper applied to the edge of the coverslip, until just sufficient pressure is applied. This may be done more precisely using a compressarium. Alternatively, methyl cellulose (Methocel) can be used. Prepare a 2% solution in water and make a ring of this on the slide. Place the liquid containing the organisms in the middle of this and cover. Immobilization occurs gradually as the methyl cellulose diffuses inwards. For the use of narcotics with protozoa see Mackinnon & Hawes (1961).

Temporary preparations

Nuclei can be demonstrated by mixing a drop of culture with a drop of 1% *methyl green* in 1% acetic acid.

Lugol's iodine (potassium iodide, 6 g; iodine 4 g; distilled water, 100 ml) stains glycogen and other substances and may also be useful for revealing nuclei and flagella.

For cilia and flagella, cirri, membranelles of ciliates etc. *Noland's method* may be used:

Phenol (saturated solution in distilled water)	80 ml
Formalin (40% HCHO)	20 ml
Glycerol	4 ml
Gentian violet	0.02 g

Moisten the stain with a little water before adding the other components. The phenol solution must not contain phenol in suspension. Add a drop of the solution to a drop of the culture.

The above methods all kill organisms. Among *vital stains* (which do not kill immediately), Janus green B is valuable for mitochondria. Use as a very dilute solution (1 in 10000 or less), or make a 0.1% solution in ethanol and allow a drop of this to dry down on a slide; add the organisms above the dried stain, cover and seal. Allow 30 min–2 h for staining to occur.

Brilliant cresyl blue

Janus green B

Nile blue A

---

methyl green (see G + S)

Hennison ?

Some ciliates and flagellates can be studied instructively by using mild detergents to dissolve the bulk of the cytoplasm, leaving behind more resistant structures such as basal bodies, fibres etc. In the original method (Child & Mazia, 1956) organisms are first fixed in 40% ethanol at $-10\ °C$ for a few hours, then transferred to cold 1% digitonin. Samples are placed on slides and covered. As the preparation warms up the cytoplasm begins to dissolve. The process can be followed under the microscope. For some organisms (e.g. large ciliates) very dilute Teepol (1 in 10^4) without preliminary fixation is equally effective.

Fixatives

Bouin's fluid (p. 23) is useful for general purposes. Fix for 15 min or more (up to 60 min for smears). Wash in several changes of 70% ethanol until the picric acid is removed.

Schaudinn's fluid is commonly used for smears. It is good for nuclei and before Feulgen.

Mercuric chloride (saturated solution in distilled water)	100 ml
Absolute ethanol	50 ml
Acetic acid (glacial)	3 ml

The acetic acid is added just before use. It may be omitted altogether. Fix for 15–60 min, then wash in 70% ethanol. Removal of mercuric chloride is facilitated by adding a few drops of Lugol's iodine (p. 84) to the ethanol while washing.

Zenker (p. 22) is useful for fixing tissues containing parasites and can also be used for smears. Fix for up to 24 h.

For better preservation of cytological detail use *Flemming*, with or without acetic (p. 25), *Champy* (p. 25) or *osmium tetroxide vapour*.

For instantaneous fixation of protozoa, giving preservation of the form of ciliary and flagellar movements, metachronal waves etc. see Párducz (1967).

Adhesion methods

For protozoa which adhere to glass (e.g. amoebae), or which live in media which act as adhesives, drop coverslips with organisms on them face down on to the surface of the fixative in a dish. Organisms which do not adhere in this way can be stuck to slides with Mayer's albumen (p. 31). Organisms are collected by centrifugation, fixed and washed in bulk and taken to 70% ethanol. A drop of the concentrate is placed on a slide smeared with albumen. The latter is coagulated by the alcohol and sticks the organisms to the slide. Individual protozoa can be handled as follows: transfer fixed organisms to a slide in a small drop of water, add a very small drop of albumen and mix thoroughly with a needle; remove most of the liquid with a fine pipette, until the organisms are just standing up out of the liquid film. Dry in a desiccator for 30 min, then place the slide in a mixture of 70% ethanol and formalin (9:1). Leave 2 min, then wash in water.

Staining methods

General methods

Heidenhain's iron haematoxylin (p. 41) is probably the most useful general method. Mallory (p. 43) and Masson's trichrome method (p. 42) are useful for parasitic protozoa in tissue sections and also for some large free-living protozoa. Use after Zenker.

Feulgen (p. 50)

This method for DNA is usually used after Schaudinn with acetic. Zenker, Flemming or Champy may also be used. The optimum hydrolysis time for Schaudinn-fixed material is about 5 min in $1 M$ HCl at 60 °C; after Champy hydrolyse for up to 25 min.

Giemsa

Used chiefly for blood parasites such as trypanosomes, but also useful for some other protozoa.

    (1) Make blood smears (p. 100) and dry by waving in air.

    (2) Fix in absolute ethanol or methanol, 3 min. Allow to dry.

(3) Float face down on Giemsa stain prepared by diluting the stock solution as purchased, 1–3 drops to 1 ml water or 0.005 M phosphate buffer, pH 7.0. Stain 20–40 min.

(4) Rinse in tap water, drain and allow to dry in air.

(5) Smears may be examined with oil-immersion, without a coverslip, and may be stored unmounted. Alternatively, mount in a neutral medium.

Chatton–Lwoff method for ciliates

This silver impregnation method (Corliss, 1953) stains cortical structures and is used particularly to demonstrate the ciliary apparatus. Organisms are fixed first in Champy, to preserve shape, and then in da Fano which is necessary before silver staining.

(1) Fix in Champy (p. 25) 1–3 min.

(2) Transfer to da Fano's fixative:

Cobalt nitrate	1 g
Sodium chloride	1 g
Formalin (40% HCHO)	10 ml
Distilled water	90 ml

This should be changed twice and material left in the third change for several hours. Material may be stored in fixative.

(3) Place a small drop of the suspension of fixed ciliates on a warm (37 °C) slide and add one drop of warm (37 °C) saline gelatin:

Gelatin	10 g
Sodium chloride	0.05 g
Distilled water	100 ml

Mix with a needle or fine glass rod and then smear to form a thin layer, slightly thicker than the ciliates. Alternatively, draw off excess fluid with a fine pipette.

(4) Cool the slide on ice or in a refrigerator until the gelatin sets (1–2 min).

(5) Transfer to 3% silver nitrate at 5–10 °C. Keep in the dark while staining. Stain 10–20 min.

(6) Rinse in ice-cold distilled water.

(7) Place face-upwards in a dish of ice-cold distilled water, 3–4 cm deep and expose to an ultraviolet lamp or bright

sunlight. Leave until the preparation is a rusty-brown colour (20–30 min).

(8) Transfer to 70% ethanol, dehydrate, clear, and mount.

The best times of staining and exposure to light are determined by trial. Single ciliates can be stained by this method.

Bodian's protargol method

Particularly useful for flagellates. It stains most cytoplasmic organelles and is especially good for Golgi bodies.

(1) Fix smears in Bouin's fluid (p. 23) or Hollande's fixative. The latter is:

Copper acetate	2.5 g
Picric acid	4 g
Formalin (40% HCHO)	10 ml
Acetic acid (glacial)	1.5 ml
Distilled water	100 ml

Dissolve the copper acetate in cold water and add the picric acid a little at a time; add the formalin and acetic acid before use.

After Bouin, wash in 70% ethanol and take smears to water. After Hollande, wash in water.

(2) Stain in 1% protargol. This should be freshly prepared. Make it up by sprinkling the protargol on the surface of distilled water in a beaker and allowing it to dissolve without stirring. Staining is carried out in Petri dishes in the presence of metallic copper (5 g to 100 ml stain), which is best added as pieces of copper wire. 10 ml of stain with 0.5 g copper is usually adequate for staining coverslip smears. Stain for 2–3 days, changing the stain on the second day. Incubation at 37 °C may be helpful.

(3) Wash in distilled water.

(4) Reduce in 1% hydroquinone in 5% sodium sulphite, 5–10 min.

(5) Wash thoroughly in distilled water.

(6) Treat with 1% gold chloride, 4–5 min.

(7) Wash in distilled water.

(8) Treat with 2% oxalic acid until the smear turns purple. This takes a few min.

(9) Wash in distilled water.

(10) Place in 5% sodium thiosulphate, 5–10 min.

(11) Wash in distilled water, dehydrate, clear and mount.

Staining of material which has been fixed in Bouin or stored for a long time may be improved by bleaching before staining. Take smears to water, treat with 5% potassium permanganate for 5 min, wash, treat with 5% oxalic acid for 5 min, wash thoroughly, then stain.

Sectioning protozoa

If protozoa are centrifuged hard enough (e.g. 700 g for 10 min) while still in fixative they will frequently form a compact pellet, which can be removed from the centrifuge tube with a small spatula and subsequently processed like a piece of tissue. Alternatively, after fixation, take organisms to water, concentrate by centrifugation and place with a fine pipette in a small cavity cut in a block of agar. Remove excess liquid and seal in the organisms with warm agar. The block is then dehydrated and sectioned as usual. Eosin or other stains may be used to make the organisms visible while sectioning.

Electron microscopy

Methods for sectioning and negative staining are essentially the same as those used for plant and animal tissues (p. 60). For sectioning, centrifugation in the fixative (see above) or the agar inclusion method may be used. Larger organisms may be processed by centrifugation at each step or handled individually with a pipette, and then sectioned individually after flat embedding (Flickinger, 1966). For scanning electron-microscope methods see Horridge & Tamm (1969) and Small & Marszalek (1969).

Insects

Whole mounts

See Eltringham (1930). Small insects or sclerotised parts of insects (mouth parts, genitalia etc.) are often best examined after macerating in 10% NaOH or KOH. Boil cautiously in this, or soak for several hours at room temperature after first immersing briefly in 95% ethanol to enable the solution to wet the material. Wash thoroughly in water, then in dilute acetic acid; dehydrate, clear and mount. Alternatively, after washing stain with Van Gieson's stain (p. 41).

Borax carmine (p. 47) may be used as a general stain for small insects or parts of insects.

When mounting, support the coverslip if necessary on strips of glass cut from slides or coverslips. Use enough mounting medium to allow for contraction as the solvent evaporates.

Chitosan test

There is no simple, specific test for chitin (see p. 54). The chitosan test (Richards, 1951) is useful but destructive.

(1) Place pieces of material, freed as far as possible from adhering tissue, in a test tube and cover with KOH solution saturated at room temperature. The tube should be closed with a Bunsen valve (i.e. a piece of rubber tubing a few cm long, sealed at one end with a glass rod or clip, with a longitudinal slit 0.5–1 cm long in its wall and connected to the tube either directly or via a rubber stopper). Heat the tube slowly on a glycerol bath to 160 °C and hold the temperature there for about 15 min.

(2) Cool to room temperature. Material remaining is likely to be chitin, converted to chitosan. Transfer to water and cover on a slide with a drop of 0.2% iodine in potassium iodide solution. Chitin turns brown. Remove excess iodine solution and replace with 1% sulphuric acid: chitin turns reddish violet.

Micro-anatomy

Rapidly penetrating fixatives are advisable. Duboscq–Brasil (p. 23) is one such; another useful mixture is picro-chlor-acetic:

Picric acid (1% in 96% ethanol)	60 ml
Chloroform	10 ml
Acetic acid (glacial)	5 ml

Fix for 12 h or longer and wash in several changes of 70% ethanol.

Difficulties in sectioning insect material may be overcome by embedding in ester wax (p. 32), agar/ester wax (p. 32), celloidin (p. 34) or epoxy-resin (p. 61).

Among standard staining methods, Mann's methyl blue/eosin (p. 45) and chlorazol black (p. 46) are recommended for staining insect material. Heidenhain's iron haematoxylin (p. 41) and the standard trichrome methods may also be used successfully.

For staining the fine musculature in partly dissected insects, use Grenacher's alum carmine (Gatenby & Beams, 1950): boil 1 g carmine in 100 ml 5% alum (aluminium potassium sulphate); cool and filter. Stain for several hours.

Tracheae and tracheoles

For demonstrating the tracheal system, including the finest tracheoles, Wigglesworth (1950) uses an injection technique. A simple apparatus is needed, consisting of a bottle with a rubber bung, connected via a three-way tap to a pump and manometer and to a reservoir of hydrogen. A glass or metal rod also passes through the bung and to the end of this is attached a glass tube, covered at the bottom with wire mesh. This can be pushed down at the appropriate stage into the injection fluid, which is in a dish placed below. The insect is placed in the tube and the pressure is reduced to 10–15 mm Hg. The bottle is then filled with hydrogen at atmospheric pressure. This is repeated once and then the pressure is reduced to 10 mm Hg again and the insect lowered into the injection fluid. Adequate preparations may also be obtained by simply reducing the pressure once and lowering the insect into the injection fluid. The latter consists of 1 part of cobalt naphthenate and 2 parts of white spirit (petroleum boiling in the range 150–190 °C).

4-2

Return to atmospheric pressure, remove the insect and wash off adherent solution as rapidly as possible by shaking with several changes of white spirit. Transfer to white spirit saturated with H_2S and bubble H_2S through the liquid for 5–60 min. Blot off excess fluid, fix in Carnoy (p. 76), and mount whole, or dissected; or embed and section. This method can be used for the study of wing tracheation.

Nervous systems

General methods

Among general histological staining methods Mallory (p. 43), Azan (p. 43) or osmic gallate (p. 48) may be used. The latter is particularly good for showing the nervous system in relation to its surroundings. Some silver impregnation methods (e.g. Nonidez, 1939; Holmes, 1947) are more specific for the nervous system, but do not succeed with all groups of animals. Special methods have been developed for groups whose nervous systems are difficult to stain: for planaria see Betchaku (1960) and Reisinger (1960); for cnidarians see Batham, Pantin & Robson (1960); for insects see Strausfeld & Blest (1970); for invertebrates generally see Fraser Rowell (1963). Methods for specific cell types in the mammalian central nervous system are described by Drury & Wallington (1967).

Methods for tracing connections in the nervous system

Staining a few cells and observing their course

These methods are capricious and stain only a small proportion of any type of cell present. The cells that do stain, however, are stained strongly and completely; the remainder are more or less unstained. This enables the stained cells to be seen in their entirety and their course to be followed.

Golgi's silver method (Colonnier, 1964). This method was devised for the light microscope but can be combined with electron microscopy (Kolb, 1970).

(1) Fix tissue by immersion or perfusion in buffered glutaraldehyde (1.75 % glutaraldehyde in 0.05 M phosphate buffer, pH 7.4, for 12 h is suitable for mammalian tissue).

(2) Transfer tissue to:

Glutaraldehyde (25 %)   20 ml

Distilled water	80 ml
Potassium dichromate	2 g

for 5–6 days at 20 °C.

(3) Transfer tissue to 0.75% silver nitrate solution for 5–6 days in the dark at 20 °C.

(4) Wash with distilled water.

(5) Embed in celloidin (p. 34).

(6) Cut thick sections (e.g. 25 μm). These can be mounted in Permount, embedding Araldite or other mounting medium for light microscopy, or can be re-embedded in Araldite and thin sectioned for electron microscopy.

If greater contrast is needed, Stell's (1965) modification may be used. Golgi's mercuric chloride method (Cox, 1891; Nauta & Ebbesson, 1970) is a useful alternative for light microscopy.

Reduced methylene blue. The McConnell (1932) method is often successful. This is used on live preparations.

(1) Add 3 drops of 25% HCl to 100 ml of 0.5% methylene blue solution. Mix thoroughly and filter.

(2) Add 2 ml of 12% Rongalite solution in distilled water to 10 ml of the filtrate.

(3) Warm carefully over a small flame. Do not boil. Stir continuously.

(4) When the deep blue solution turns deep dirty green, remove from the heat and continue stirring.

(5) Stop stirring when the solution is almost clear, with a yellow precipitate. Filter when cold.

(6) Allow to stand for 24–36 h before use. The solution will keep for 8–10 days.

(7) Add 1–2 ml of this stain to 25 ml of isotonic saline and immerse the preparation in it. The solution will turn light, milky blue, then dark blue.

(8) Stain for a few seconds to 2 h.

(9) After staining, keep the preparation in a well-oxygenated situation so that the dye can be reoxidised by the tissue.

(10) Observe on a slide under a coverslip.

The preparation may be made permanent as follows. The first five steps should be performed at 0 °C.

(1) Fix for 8 min in saturated ammonium picrate solution in isotonic saline.

(2) Transfer to 8% ammonium molybdate solution for several hours.

(3) Wash in tap water.

(4) Dehydrate in the following series:

	water	95% ethanol	n-butanol
(i)	43	30	27
(ii)	30	30	40
(iii)	18	27	55
(iv)	9	21	70
(v)	3	12	85
(vi)	0	0	100

Allow 20 min in each mixture.

(5) Transfer to equal parts of n-butanol and methyl benzoate for 20 min, then to pure methyl benzoate.

(6) Allow to warm up to room temperature and transfer to toluene.

(7) Mount in balsam, or embed in paraffin and section in the usual way.

Staining myelin

Pal's (1886) modification of Weigert's method may be used.

(1) Fix thick slices of tissue in:

Potassium dichromate	2.5 g
Sodium sulphate	1 g
Distilled water	100 ml

(2) Wash in water.

(3) Embed in celloidin (p. 34) and cut 25 μm sections. Stain these loose.

(4) Wash sections in distilled water.

(5) Stain for 2–3 days at room temperature in:

Haematoxylin (10% in absolute ethanol, ripened)	10 ml
Distilled water	90 ml

Lithium carbonate (saturated aqueous solution,
    approx. 1.3%)   1 ml
(6) Rinse in tap water.
(7) Treat with 0.25% potassium permanganate for 30 s.
(8) Rinse in tap water.
(9) Differentiate in Pal's bleach:

Oxalic acid (1%)   50 ml
Potassium sulphite (1%)   50 ml

    Mix immediately before use.

It is best to carry out stages (7), (8) and (9) cyclically, with a few sec in (7) and 1 min in (9) until the grey matter is white and the white matter blue-grey.

(10) Wash thoroughly in tap water.
(11) If the tap water is not alkaline place for 5 min in:

Lithium carbonate (saturated aqueous solution)   7 ml
Distilled water   93 ml

    Then rinse in tap water.
(12) Dehydrate, clear and mount.

This method is not ideal for tracing connections in the nervous system. Myelinated axons often lose their myelin before they terminate, and they often interweave confusingly. Non-myelinated axons do not stain. The method can, however, be used in conjunction with degeneration experiments: fully degenerated myelin does not stain, so degenerated tracts can be followed by serial sectioning.

Other methods of staining myelin may be found in Drury & Wallington (1967).

Degeneration studies

The effects of cutting nerves may be detectable histologically from 3 days to many weeks after the operation. The best degeneration time for each technique and each animal has to be determined by experiment. Note that in the U.K. such experiments need a Home Office licence if performed on vertebrates.

Gallocyanin method for degenerating cell bodies. Cell bodies whose axons have been cut can be identified since they slowly undergo retrograde chromatolysis: that is, they swell, their nuclei become excentrically placed and, in mammals, they lose their rough endoplasmic reticulum (Nissl substance). This may be shown up by staining with gallocyanin (Berube, Powers, Kerkay & Clark, 1966).

Gallocyanin	150 mg
Chrome alum	15 g
Distilled water	100 ml

Boil these together for 20 min, allow to cool and filter. Wash the residue with distilled water until the total filtrate is 100 ml. Adjust the pH of the filtrate to 8.5 with dilute ammonia. Filter off the precipitate, using a filter pump, and wash with a small volume of anhydrous ether. Dry, and store in a sealed container in the deep freeze. For use, make up a 3% solution in $1\,\mathrm{M}\,H_2SO_4$.

Use material fixed in formaldehyde solutions (p. 21) or acetic-ethanol (p. 76).

(1) Bring sections to water.

(2) Stain with gallocyanin, 24–48 h.

(3) Rinse with distilled water brought to pH 1.6 with HCl.

(4) Repeat until the rinse is colourless.

(5) Dehydrate, clear and mount.

Normal cell bodies will have their Nissl substance stained dark blue. Cell bodies whose axons have been cut for a week or more will be palely and diffusely stained. Very occasionally not only the damaged neuron shows this phenomenon but also those with which it synapses. Insect neurons react rather differently to cutting: after 48 h they develop a perinuclear ring of RNA. The presence of this may be used to trace connections. As in mammals, the nucleus moves to an excentric position after a few weeks (Cohen & Jacklet, 1967).

Staining degenerating myelin. The degenerating myelin on axons that have been cut can be stained positively by the Swank & Davenport (1935) modification of the Marchi method. Care is necessary to avoid tissue damage, both before and after fixation, as this may cause artifacts.

(1) Fix slices of tissue in formol-saline (p. 22), 2–7 days.

(2) Place fixed tissue directly in:

Potassium chlorate (1%)	60 ml
Osmium tetroxide (1%)	20 ml
Acetic acid (glacial)	1 ml
Formalin (40% HCHO)	12 ml

(Poirier, Ayotte & Gautier (1954) use a mixture containing the same ingredients but at substantially lower concentrations, which is equally satisfactory.) Support the tissue on a layer of glass wool and handle carefully. Agitate periodically. Leave for 7–10 days.

(3) Wash in running water, 24 h.

(4) Dehydrate, embed in celloidin (p. 34), section and mount. Degenerating myelin is stained black and can be traced in serial sections. Controls for stainable substances other than degenerating myelin should be performed by treating tissue from intact animals by the same technique. The method has the disadvantages of the Weigert procedure (p. 95) but does produce a positive picture.

Silver impregnation methods. Degenerating axons may be impregnated by the Nauta method (Nauta, 1957; Nauta & Ebbesson, 1970). The synapses of axons that have been cut may be shown up in certain regions of the mammalian central nervous system a few days after the operation by Glees' (1946) silver method. This may depend on staining the large number of neurofilaments that occur in some regenerating synapses (Guillery, 1965).

Dye injection methods

Dyes may be injected into cells by pressure from a syringe or iontophoretically by passing current through a dye-filled glass micro-electrode. Large molecular weight, fluorescent, Procion dyes (I.C.I. Ltd; distributed by Dylon International Ltd, Dylon Works, Sydenham, London) are normally used (Stretton & Kravitz, 1968). If the cell is left for 15–96 h after injection the entire cell will be filled with dye. The material can then be fixed with 6% glutaraldehyde at pH 4.0 and the cell found in frozen or paraffin sections by use of the fluorescence microscope (p. 5). The Procion dyes are usually confined to the cell injected.

Injected dyes such as fast green F.C.F. may also be used, in conjunction with electrophysiology, to localise histologically the site of a micro-electrode (Erulkar, Nichols, Popp & Koelle, 1968).

Methods for identifying synaptic transmitters

These methods aim at histochemical demonstration of the localisation of transmitter or associated substances, or of the enzymes that form or break down the transmitter. For fluorescence studies of catecholamines see Marsden & Kerkut (1969), who used the method of Falck, Hillarp, Thieme & Torp (1962); see also Rude, Coggeshall & van Orden (1969). An immunofluorescent method for localising protein associated with catecholamines is described by Geffen, Livett & Rush (1969). For fluorescence studies of histamine see Ehinger & Thunberg (1967).

Catecholamines may be revealed in the electron microscope by the methods of Woods (1969) or Wood & Barrnett (1964).

For demonstration of acetylcholinesterase at the electron-microscope level see Lewis & Shute (1966); the method may be used at the light-microscope level by diluting the incubation medium with an equal volume of isotonic sodium sulphate solution and incubating for at least 12 h at room temperature.

Blood smears

Blood is usually examined in the light microscope as a smear that has been dried rapidly in air and then fixed and stained. In order to obtain a lifelike appearance and uniform distribution of the cells, the blood should be smeared and dried as quickly as possible.

Making the smear

(1) Use clean, grease-free microscope slides (p. 116). Handle these only by their edges.

(2) Sterilise the skin at the base of the thumb nail, or the tip of a finger, by rubbing vigorously with a pad soaked in 70% alcohol. Allow the skin to dry.

(3) Pierce the skin by a quick jab with a sterile lancet. Do not

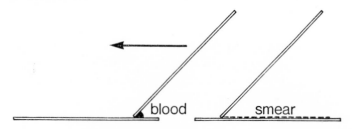

Figure 3. Diagram illustrating the method of making a blood smear.

squeeze near the puncture for this will alter the blood count. To increase the amount of blood produced, tightly encircle the base of the punctured thumb or finger with the other hand, then slowly bend the digit that is grasped. Discard the first drop of blood.

(4) Within 5 s touch the end of a slide onto the second drop of blood.

(5) Hold this slide at 45° to the horizontal, with the drop of blood downwards, and touch it onto one end of a second grease-free slide placed horizontally on the bench (figure 3).

Notice that the held slide makes an acute angle with the horizontal slide. The blood should spread across the width of the

100

slides by capillarity. As this happens, move the held slide rapidly along the length of the horizontal slide so that the blood is drawn along *behind* the held slide. Do not move the slide in such a way that it is pushed *over* the drop of blood. This whole operation should take less than one second. The angle of the slides and the speed of movement influence the thickness of the smear; a slow movement and a large angle both produce a thick smear. Hesitation at the beginning of the movement allows white cells to adhere to the glass, so the remainder of the smear is depleted.

(6) Dry the film rapidly by waving the slide in air.

Fixing and staining

A combined fixative and stain is used, prepared by adding 0.15 g of Leishman's stain (powder) to 100 ml of methanol. Shake occasionally. The stain is ready for use after about 24 h.

(1) With a lump of paraffin wax draw a line across the slide at each end of the smear. This helps to contain the stain.

(2) Add enough Leishman's stain to cover the smear (5 drops) and allow 3 min for the methanol to fix the cells. Minimise evaporation by inverting a large watchglass over the slide. Evaporation causes precipitation of insoluble granules of stain.

(3) Dilute the stain with 10 drops of phosphate buffer (pH 7.2, p. 113), mix thoroughly, and leave for 10 min.

(4) Drain off the stain and wash the smear in phosphate buffer or glass-distilled water until there is no tinge of blue or grey in the smear as seen with the naked eye.

(5) Drain off as much water as possible and allow the smear to dry thoroughly in a warm place.

(6) Examine unmounted, using an oil-immersion objective, or add one or two drops of Canada balsam and cover with a long coverslip.

NOTE: *Bone marrow* smears may be processed in the same way but need to be stained for 18–20 min and will remain blue even when thoroughly washed (see Drury & Wallington, 1967).

Manipulation of blood

Any glass surface (except slides for smears) that will come into contact with blood should be siliconised by immersing in a solution of silicone in carbon tetrachloride which is then allowed to evaporate; alternatively the glassware may be rinsed in a water-soluble silicone concentrate (for example 'Siliclad' obtainable from Clay–Adams, Inc. New York, N.Y. 10010, U.S.A., or from Arnold Horwell Ltd, 2 Grangeway, Kilburn High Road, London) and dried in an oven. This will reduce the speed of clotting.

Prevention of clotting

(1) By mixing the blood with glass beads. This removes the fibrin and the platelets. Three glass beads for each 10 ml of blood are added to the freshly drawn blood in a stoppered tube or round-bottomed flask. The tube is repeatedly inverted to mix the beads and blood for a continuous period of 3 min; if the flask is used, it should be rotated to swirl the beads in the blood for 10 min. The fibrin and platelets form a spongy mass that can be removed. This process destroys about 25% of the red cells, but the white cells are healthy.

(2) By the use of heparin. Heparin is usually obtained as a powder containing 100 i.u. mg^{-1} of heparin. It should be mixed with the freshly drawn blood at the rate of 0.02–0.1 mg ml^{-1} blood. The motility and phagocytic activity of leucocytes is not impaired by heparin but these cells may clump abnormally.

(3) By chelation of the calcium ions necessary for clotting. Add sodium citrate solution (0.4 ml of 20% solution to 9.6 ml of blood) or EDTA (4.5 ml of 2.5% disodium ethylene diamine tetra-acetic acid added to 75.0 ml blood). Sodium oxalate is sometimes used but is toxic. Chelating agents damage blood platelets.

Separation of blood cells

Red and white cells may be separated by centrifuging blood at 200 g for 5 min. The red cells sediment first and the white cells form the whitish 'buffy coat' between the loose pellet of red cells and the plasma. Platelets remain in suspension. The 'buffy coat' may be removed and the red cells resuspended with a pipette. Repeat twice more for a good separation of cell types.

Alternatively the separation may be performed by adding to the blood an equal volume of 6% dextran (M.W. 139 000 to 228 000) in 0.9% saline.[1] The dextran causes the red cells of mammals other than ungulates and rats to form rouleaux and to precipitate under the influence of gravity, leaving the white cells in suspension. Precipitation of the red cells is complete in 45 min. White cells may be obtained by removing the suspension and centrifuging it at 200 g for 5 min. Platelets remain in suspension.

Further methods for separating blood cells are given in Cutts (1970).

[1] Marketed as a sterile solution, 'Dextraven 110', by Fisons Pharmaceuticals Ltd, Loughborough, Leics. The molecular weight of the dextran in this solution is slightly less than optimum, but the solution is effective.

Saline media

General notes

The ideal saline medium is one in which tissues behave as if they were in the organism. It is often simplest and most satisfactory to use blood or tissue fluid as a saline. Some blood, however, such as that of insects and the snail *Viviparus*, autolyses on contact with air. The blackening and development of toxicity of insect blood on exposure to air can be prevented (Jones & Cunningham, 1961). Collect blood in a tube chilled in ice to stop clotting. Heat the blood to 60 °C for 5 min. Place in a deep freeze for 12 h. Allow to thaw and centrifuge at 6000 g for 10 min. Use the clear supernatant.

Liquid paraffin is a useful alternative to a saline for some purposes, since it prevents the evaporation of water. Oxygen and carbon dioxide are quite soluble in this: liquid paraffin dissolves approximately its own volume of CO_2 and 0.15 its volume of oxygen at NTP. Approximately 80 parts of water dissolve in a million parts of liquid paraffin.

Sea water (see below) is for many purposes a fair saline medium for the tissues of marine organisms.

For simple, non-critical work the salines listed in Table 2 are adequate. Add the salts in the order given in the table. The quantities are given as grams of the anhydrous salt per litre. In some of the original formulae (e.g. Tyrode's solution) the state of hydration of the salt used is not stated. Where a common hydrate exists (e.g. $MgCl_2.6H_2O$) it has been assumed that the author used this; where there is no one common hydrate (e.g. $CaCl_2$) it has been assumed that the anhydrous salt was used.

All saline media should have air bubbled through them before and, ideally, during use.

For more critical work the salines must be matched more closely to the particular species or the particular tissue, or the particular property it is desirable to preserve. This is especially true of insects, the haemolymph of which varies very widely in composition from species to species. Lockwood (1961) lists the composition of saline solutions for many common animals.

TABLE 2. *Composition of saline solutions for various organisms*

(Unless otherwise stated, the quantities given are the weight in grams of the anhydrous salts contained in one litre of solution.)

	NaCl	KCl	CaCl$_2$	NaHCO$_3$	NaH$_2$PO$_4$	MgCl$_2$	Glucose	Notes
Earthworm (Roots, 1955)	4.1	0.082	0.084	0.28	—	—	—	—
Helix (Kerkut & Laverack, 1957)	8.6	0.21	0.34	—	—	0.13	—	Add 0.19 g KH$_2$PO$_4$
Crustacea, marine (Hagiwara & Bullock, 1957)	26.4	1.12	2.78	—	—	0.178	—	Add 0.48 g MgSO$_4$.7H$_2$O and 0.55 g H$_3$BO$_3$. Adjust pH to 7.5 with NaOH
Insect saline	7.5	—	—	—	—	—	—	—
Cockroach Ringer (Yamasaki & Narahashi, 1959)	12.28	0.23	0.20	—	0.03	—	—	Add 0.25 g Na$_2$HPO$_4$
Diptera Ringer (Berridge, unpublished)	7.5	1.5	0.22	—	0.80	0.192	1.8	Add 0.5 g Na glutamate, 0.37 g malic acid, 0.37 g citric acid and sufficient NaOH (approx. 30 pellets) to bring the pH to 7.0
Locust Ringer (see p. 105)								

Locust Ringer (Weis-Fogh, 1956)

NaCl (29.2 g in 1000 ml)	210 ml
KCl (37.3 g in 1000 ml)	20 ml
$NaH_2PO_4.2H_2O$ (78.0 g in 1000 ml)	30 ml
$Na_2HPO_4.12H_2O$ (89.5 g in 1000 ml)	70 ml
$CaCl_2.6H_2O$ (21.9 g in 1000 ml)	20 ml
$MgCl_2.6H_2O$ (20.3 g in 1000 ml)	20 ml

Distilled water to make 1000 ml.
Penicillin (30 mg l^{-1}) may be added.

Mix the solutions in the order given. The pH should be 6.8.

Culture solutions for flowering plants

Pfeffer

$Ca(NO_3)_2.4H_2O$	4 g
KNO_3	1 g
$MgSO_4.7H_2O$	1 g
KH_2PO_4	1 g
KCl	5 g
$FeCl_3$	0.1 g

Distilled water to make 3–7 litres

For *Knop's* solution omit the KCl and make up as above b
12 l distilled water. For solutions deficient in one element r
Ca^{2+} by K^+, K^+ by Ca^{2+}, Mg^{2+} by K^+, NO_3^- by Cl^-, I
SO_4^{2-}; omit $FeCl_3$.

Culture solutions for bacteria, algae and fungi

See Norris & Ribbons (1969, 1970) and Booth (197

Sea water

Salinity is expressed as grams per kilogram
per mille, ‰). This is measured by titrating
water (*chlorinity*, also expressed as gram
water) with silver nitrate solution.

								Buffer with phosphate or bicarbonate
Teleost, freshwater (Young, 1933)	5.5	0.14	0.12	—	—	—	—	—
Teleost, marine (Forster & Hong, 1958)	7.8	0.18	0.166	0.084	0.06	0.095	—	—
Amphibian saline	6.0	—	—	—	—	—	—	—
Amphibian saline (Boyle & Conway, 1941)	4.24	0.148	—	2.1	—	—	4.684	Add 0.356 g Na_2HPO_4, 0.146 g $MgSO_4$, 0.092 g Na_2SO_4, 0.068 g KH_2PO_4, 0.4 g Ca gluconate. Bubble with 97 % $O_2 + 3$ % CO_2
Amphibian embryo (Holtfreter, 1943)	3.5	0.05	0.1	0.2	—	—	—	—
Chick embryo (Ringer)	9.0	0.42	0.24	—	—	—	—	—
Mammalian saline	0.9	—	—	—	—	—	—	—
*Mammalian Ringer (Earle, 1943)	6.8	0.4	0.2	2.2	0.11	—	1.0	Add 0.1 g $MgSO_4$; sterilise by filtration
*Tyrode (1910)	8.0	0.2	0.2	1.0	0.04	0.05	1.0	Make up immediately before use. Adjust pH with 0.1 M HCl
Plant material (Ringer)	7.5	0.075	0.1	0.1	—	—	—	—

NOTE: The pH of simple salt solutions may be low; that of other solutions (e.g. Tyrode) may be high. For many purposes it is satisfactory to adjust the pH to 7.4.

* Add the salts to one litre of water.

Chlorosity is expressed as grams per litre.

Molarity of Cl in sea water =

$$\frac{(\text{vol. of AgNO}_3)\,(\text{Molarity of AgNO}_3)}{\text{vol. of sea water}}$$

Chlorosity $\qquad = 35.5 \,(\text{Molarity of Cl})$

Chlorinity $\qquad = \dfrac{\text{Chlorosity}}{\text{Density of sample at 20 }^\circ\text{C}}$

Properties of North Sea water are as follows: salinity, 34.325‰; chlorinity, 19‰; freezing point, −1.872 °C; density 1.0245 g ml^{-1} at 20 °C.

Solutions isotonic with sea water

	Molarity	g l^{-1} solution
NaCl	0.552	32.27
KCl	0.562	41.90
$CaCl_2.6H_2O$	0.380	83.18
$MgCl_2.6H_2O$	0.370	75.37
$Na_2SO_4.10H_2O$	0.489	157.76
$MgSO_4.7H_2O$	0.867	213.73
$NaHCO_3$	0.575	48.35
KBr	0.552	65.74
Sucrose	0.783	267.95
Urea	0.999	60.00

Note that these solutions vary widely in pH.

Artificial sea water

0.55 M NaCl	747.8 ml
0.56 M KCl	16.3 ml
0.38 M $CaCl_2$	26.8 ml
0.37 M $MgCl_2$	145.03 ml
0.49 M Na_2SO_4	58.21 ml
0.55 M KBr	1.5 ml
$NaHCO_3$	0.197 g
Distilled water	to 1000 ml

Data from Nicol (1960). See Nicol (1960) for the composition of sea water at different dilutions.

Sea water has a pH of 8.1 and is quite well buffered. The pH may rise to 9.6 in pools owing to the activity of plants, and in aquaria it may fall to 7, due to decomposition and the lack of plants. The pH is kept high in some aquaria by adding bicarbonate or lime; this will affect the composition of the sea water.

Buffers

See Gomori (1955), Diem (1962). For the osmotic pressure of common buffers see Maser, Powell & Philpott (1968).

Acetic acid/acetate buffer (Walpole)

A. 0.2 M acetic acid (11.55 ml in one litre)
B. 0.2 M sodium acetate (27.2 g $CH_3COONa.3H_2O$ in one litre)

Add the volumes of A and B given below and make up to 100 ml with distilled water.

ml A	ml B	pH
46.3	3.7	3.6
44.0	6.0	3.8
41.0	9.0	4.0
36.8	13.2	4.2
30.5	19.5	4.4
25.5	24.5	4.6
20.0	30.0	4.8
14.8	35.2	5.0
10.5	39.5	5.2
8.8	41.2	5.4
4.8	45.2	5.6

Cacodylate buffer (Plumel)

Commonly used for buffering aldehydes when used for fixation in electron microscopy. CAUTION: contains arsenic.

A. 0.2 M sodium cacodylate (42.8 g $Na(CH_3)_2AsO_2.3H_2O$ in one litre)
B. 0.2 M HCl

Add the volume of B shown below to 50 ml of A and make up to 200 ml with distilled water.

ml B	pH	ml B	pH
47.0	5.0	18.3	6.4
45.0	5.2	13.3	6.6
43.0	5.4	9.3	6.8
39.2	5.6	6.3	7.0
34.8	5.8	4.2	7.2
29.6	6.0	2.7	7.4
23.8	6.2		

Citric acid/phosphate buffer (McIlvaine)

A. 0.1 M citric acid (21.0 g in one litre)
B. 0.2 M disodium hydrogen phosphate
   (35.6 g $Na_2HPO_4.2H_2O$ in one litre)

To the volume of B shown below add sufficient A to make 100 ml.

ml B	pH	ml B	pH
2.0	2.2	53.6	5.2
6.2	2.4	55.8	5.4
10.9	2.6	58.0	5.6
15.8	2.8	60.5	5.8
20.6	3.0	63.2	6.0
24.7	3.2	66.1	6.2
28.5	3.4	69.3	6.4
32.2	3.6	72.8	6.6
35.5	3.8	77.2	6.8
38.6	4.0	82.4	7.0
41.4	4.2	86.9	7.2
44.1	4.4	90.7	7.4
46.8	4.6	93.6	7.6
49.3	4.8	95.7	7.8
51.5	5.0	97.2	8.0

Glycine/HCl buffer (Sørensen)

Glycine (0.1 M)   7.50 g
Sodium chloride (0.1 M)   5.85 g
Distilled water to make one litre.

To the volume of 0.1 M HCl indicated below add sufficient of the stock solution to make 100 ml.

ml HCl	pH	ml HCl	pH	ml HCl	pH	ml HCl	pH
94.3	1.1	58.3	1.7	39.7	2.3	20.4	2.9
85.4	1.2	54.7	1.8	36.4	2.4	17.9	3.0
77.4	1.3	51.1	1.9	33.4	2.5	15.2	3.1
71.1	1.4	48.1	2.0	30.4	2.6	12.0	3.2
66.2	1.5	45.1	2.1	27.2	2.7	10.8	3.3
62.0	1.6	42.4	2.2	24.0	2.8		

Glycine/NaOH buffer (Sørensen–Walbum)

Glycine (0.1 M)   7.50 g
Sodium chloride (0.1 M)   5.85 g
Distilled water to make one litre.

To the volume of 0.1 M NaOH indicated below add sufficient of the stock solution to make 100 ml.

ml NaOH	pH	ml NaOH	pH
5	8.45	50	11.14
10	8.79	51	11.39
20	9.22	55	11.92
30	9.56	60	12.21
40	9.98	70	12.48
45	10.32	80	12.66
49	10.90	90	12.77

Phosphate buffer (Sørensen)

A. M/15 potassium dihydrogen phosphate
(9.08 g KH_2PO_4 in one litre)
B. M/15 disodium hydrogen phosphate
(11.88 g $Na_2HPO_4.2H_2O$ in one litre)

To the volume of B shown below add sufficient A to make 100 ml.

ml B	pH	ml B	pH
1.2	5.0	49.2	6.8
2.0	5.2	60.8	7.0
3.3	5.4	71.5	7.2
5.0	5.6	80.4	7.4
8.1	5.8	86.8	7.6
12.3	6.0	91.4	7.8
18.5	6.2	94.5	8.0
26.8	6.4	96.7	8.2
37.3	6.6		

TRIS buffer (Gomori)

Cannot be used for buffering aldehydes.

A. 0.2M tris (hydroxymethyl) aminomethane
(24.2 g in one litre)
B. 0.2M HCl

To the volume of B shown below add 50 ml of A and make up to 200 ml with distilled water.

ml B	pH	ml B	pH
44.2	7.2	21.9	8.2
41.4	7.4	16.5	8.4
38.4	7.6	12.2	8.6
32.5	7.8	8.1	8.8
26.8	8.0	5.0	9.0

Veronal-acetate buffer (Michaelis)

Cannot be used to buffer aldehydes.

A. Sodium acetate ($CH_3COONa.3H_2O$) 9.714 g
Sodium veronal (sodium barbitone, sodium diethyl barbiturate) 14.714 g
Distilled water to make 500 ml.
B. 0.1 M HCl

Add the volume of B shown below to 20 ml A and make up to 100 ml with distilled water.

ml B	pH	ml B	pH
64	2.62	24	6.99
60	3.20	22	7.25
56	3.62	20	7.42
52	3.88	16	7.66
48	4.13	12	7.90
44	4.33	8	8.18
40	4.66	4	8.55
36	4.93	3	8.68
32	5.32	2	8.90
28	6.12	1	9.16
26	6.75		

SI units

The Système International d'Unités (SI units) is coming into increasing use in biological publications and will probably soon become standard, at least in Europe. It is an extension and refinement of the existing metric system. From the point of view of microscopy, the two important features of SI units are, first, that the basic units of length, mass and time are the metre (m), kilogram (kg) and second (s), and secondly that fractions of the basic units should normally be restricted to steps of a thousandth. The fractions most likely to be used in microscopy are as follows:

fraction	prefix	symbol
10^{-3}	milli	m
10^{-6}	micro	μ
10^{-9}	nano	n

Thus, 10^{-9} metre is written as 1 nm, and 10^{-6} metre as 1 μm. The ångström unit (Å, = 10^{-10} m), though much used by crystallographers and electron microscopists, does not form part of the SI system and it is proposed that its use should be progressively abandoned.

For further details see:

British Standards Institution (1969). *The Use of SI Units.* Publication PD 5686. London: BSI.

Royal Society (1971). *Quantities, Units and Symbols.* London: Royal Society.

Cleaning glassware

The method to be used depends on what the glassware is to be used for. Great care is needed with glassware to be used for living material. Chromic acid cleaning mixture and most detergents are extremely difficult to wash off. The surfactant RBS 25 (Chemical Concentrates Ltd, 41 Webb's Road, London SW 11) is a seemingly safe, non-toxic cleaning agent. For general purposes soak glassware for 24 h in a 2% solution and rinse thoroughly in water. Glassware to be used for culture work or living material should be soaked in a more dilute solution of RBS 25 (0.25%) and again rinsed very thoroughly. Paul (1970) describes other methods for cleaning glassware to be used for tissue culture work.

Chromic acid cleaning mixture will remove most kinds of contamination but is unpleasant to handle and should be used only when essential. Prepare by adding cautiously 1000 ml concentrated H_2SO_4 to 35 ml of a saturated solution of sodium dichromate in water. Soak glassware for several hours, then rinse very thoroughly. (If the glassware is to be used for living material it is recommended to rinse in 9 changes of tap water and 3 of distilled water.) Discard chromic acid solutions with a greenish colour.

An ultrasonic cleaning bath may be helpful in dislodging tenacious dirt deposits or in cleaning objects of complicated shape.

New slides and coverslips for ordinary histological work may be cleaned by dipping in 70% ethanol and drying with a cloth.

List of Colour Index numbers of stains

Acid fuchsin	42685
Aniline blue (W.S.)	42755
Azocarmine B	50090
Azocarmine G	50085
Azure A	52005
Basic fuchsin	42510
Biebrich scarlet	26905
Chlorazol black E	30235
Cotton blue (methyl blue)	42780
Crystal violet	42555
Eosin B	45400
Eosin Y	45380
Fast green (F.C.F.)	42053
Light green	42095
Methyl blue (cotton blue)	42780
Methyl green	42585
Oil red O	26125
Orange G	16230
Ponceau 2R	16150
Ponceau S	27195
Pyronin Y (G)	45005
Resorcin blue	51400
Safranin O	50240
Sudan black B	26150

Bibliography

ALLEN, R. M. (1958). *Photomicrography*, 2nd edn. Princeton: Van Nostrand.

ASHBURNER, M. (1967). Patterns of puffing activity in the salivary gland chromosomes of *Drosophila*. 1. Autosomal puffing patterns in a laboratory stock of *Drosophila melanogaster*. *Chromosoma* **21**, 398–428.

BAKER, J. R. (1944). The structure and chemical composition of the Golgi element. *Q. Jl microsc. Sci.* **85**, 1–71.

BAKER, J. R. (1947). The histochemical recognition of certain guanidine derivatives. *Q. Jl microsc. Sci.* **88**, 115–21.

BAKER, J. R. (1956). The histochemical recognition of phenols, especially tyrosine. *Q. Jl microsc. Sci.* **97**, 161–4.

BAKER, J. R. (1958). *Principles of Biological Microtechnique: a Study of Fixation and Dyeing.* London: Methuen.

BAKER, J. R. (1960). *Cytological Technique*, 4th edn. London: Methuen.

BARER, R. (1966). Phase contrast and interference microscopy in cytology. In *Physical Techniques in Biological Research*, 2nd edn, vol. 3A (ed. A. W. Pollister), pp. 1–56. New York and London: Academic Press.

BARER, R. (1968). *Lecture notes on the Use of the Microscope*, 3rd edn. Oxford: Blackwell.

BATHAM, E. J., PANTIN, C. F. A. & ROBSON, E. A. (1960). The nerve-net of the sea-anemone *Metridium senile*: the mesenteries and the column. *Q. Jl microsc. Sci.* **101**, 487–510.

BELL, G. R. (1964). A guide to the properties, characteristics, and uses of some general anaesthetics for fish. *Bull. Fish. Res. Bd Can.* **148**, 1–5.

BENJAMINSON, M. A. (1969). Conjugates of chitinase with fluorescein isothiocyanate or lissamine rhodamine as specific stains for chitin *in situ*. *Stain Technol.* **44**, 27–31.

BENNETT, H. S. (1950). The microscopical investigation of biological materials with polarized light. In *McClung's Handbook of Microscopical Technique*, 3rd edn (ed R. McClung Jones). New York: Hoeber.

BERGERON, J. A. & SINGER, M. (1958). Metachromasy: an experimental and theoretical reevaluation. *J. biophys. biochem. Cytol.* **4**, 433–57.

BERUBE, G. R., POWERS, M. M., KERKAY, J. & CLARK, G. (1966). The gallocyanin–chrome alum stain; influence of methods of preparation on its activity and separation of active staining compound. *Stain Technol.* **41**, 73–81.

BETCHAKU, T. (1960). A copper sulfate–silver nitrate method for nerve fibers of planarians. *Stain Technol.* **35**, 215–8.

BONE, Q. & DENTON, E. J. (1971). The osmotic effects of electron microscope fixatives. *J. Cell Biol.* **49**, 571–81.

BOOTH, C. (ed.) (1971). *Methods in Microbiology*, vol. 4. London and New York: Academic Press.

BOYLE, P. J. & CONWAY, E. J. (1941). Potassium accumulation in muscle and associated changes. *J. Physiol., Lond.* **100**, 1–63.

BRADLEY, D. E. (1965). Replica and shadowing techniques. In *Techniques for Electron Microscopy* (ed. D. H. Kay), 2nd edn, pp. 96–152. Oxford: Blackwell.

BRAIN, E. B. (1966). *The Preparation of Decalcified Sections*. Springfield, Illinois: Thomas.

BURGOS, M. H., VITALE-CALPE, R. & TÉLLEZ DE IÑON, M. T. (1967). Studies on paraformaldehyde fixation for electron microscopy. 1. Effect of concentration on ultrastructure. *J. Microscopie* 6, 457–68.

BURSTONE, M. S. (1962). *Enzyme Histochemistry*. New York and London: Academic Press.

CANNON, H. G. (1936). *A Method of Illustration for Zoological Papers*. Association of British Zoologists.

CARO, L. G. (1964). High resolution autoradiography. In *Methods in Cell Physiology*, vol. 1 (ed. D. M. Prescott), pp. 327–63. New York and London: Academic Press.

CASON, J. E. (1950). A rapid one-step Mallory–Heidenhain stain for connective tissue. *Stain Technol.* 25, 225–6.

CHAMBERS, R. & CHAMBERS, E. L. (1961). *Explorations into the Nature of the Living Cell*, pp. 166–74. Cambridge, Mass.: Harvard University Press.

CHIFFELLE, T. L. & PUTT, F. A. (1951). Propylene and ethylene glycol as solvents for Sudan IV and Sudan Black B. *Stain Technol.* 26, 51–6.

CHILD, F. M. & MAZIA, D. (1956). A method for the isolation of the parts of ciliates. *Experientia* 12, 161–2.

COHEN, M. J. & JACKLET, J. W. (1967). The functional organisation of motor neurons in an insect ganglion. *Phil. Trans. R. Soc. B.* 252, 561–9.

COLONNIER, M. (1964). The tangential organisation of the visual cortex. *J. Anat.* 98, 327–44.

CONN, H. J. (1961). *Biological Stains*, 7th edn. Baltimore, Maryland: Williams & Wilkins.

CONN, H. J., DARROW, M. A. & EMMEL, V. M. (1960). *Staining procedures used by the Biological Stain Commission*, 2nd edn. Baltimore, Maryland: Williams & Wilkins.

COONS, A. H. (1958). Fluorescent antibody methods. In *General Cyto-chemical Methods*, vol. 1 (ed. J. F. Danielli), pp. 399–422. New York: Academic Press.

CORLISS, J. O. (1953). Silver impregnation of ciliated Protozoa by the Chatton–Lwoff technic. *Stain Technol.* 28, 97–100.

COX, W. H. (1891). Imprägnation des centralen Nervensystems mit Quecksilbersalzen. *Arch. mikrosk. Anat.* 37, 16–21.

CRUICKSHANK, R. (ed.) (1969). *Medical Microbiology*, 11th edn. Edinburgh and London: Livingstone.

CURTIS, A. S. G. (1960). Area and volume measurements by random sampling methods. *Med. biol. Illust.* 10, 261–6.

CUTTS, J. H. (1970). *Cell Separation Methods in Haematology*. New York and London: Academic Press.

DIEM, K. (ed.) (1962). *Documenta Geigy Scientific Tables*, 6th edn. Macclesfield: Geigy (U.K.) Ltd.

119

BIBLIOGRAPHY

DRURY, R. A. B. & WALLINGTON, E. A. (1967). *Carleton's Histological Technique*, 4th edn. New York and Toronto: Oxford University Press.

EARLE, W. R. (1943). Production of malignancy *in vitro*. IV. The mouse fibroblast cultures and changes seen in the living cells. *J. natn. Cancer. Inst.* **4**, 165–212.

ECHLIN, P. (1971*a*). The application of scanning electron microscopy to biological research. *Phil. Trans. R. Soc.* B **261**, 51–9.

ECHLIN, P. (1971*b*). Preparation of labile biological material for examination in the scanning electron microscope. In *Scanning Electron Microscopy* (ed. V. H. Heywood). New York and London: Academic Press.

EGGEN, R. R. (1965). *Chromosome Diagnostics in Clinical Medicine*. Springfield, Illinois: Thomas.

EHINGER, B. & THUNBERG, R. (1967). Induction of fluorescence in histamine-containing cells. *Expl. Cell Res.* **47**, 116–22.

ELTRINGHAM, H. (1930). *Histological and Illustrative Methods for Entomologists*. Oxford: Clarendon Press.

ENGEL, C. E. (ed.) (1968). *Photography for the Scientist*. New York and London: Academic Press.

ERÄNKÖ, O. (1955). *Quantitative Methods in Histology and Microscopic Histochemistry*. Basel and New York: Karger.

ERULKAR, S. D., NICHOLS, C. W., POPP, M. B. & KOELLE, G. B. (1968). Renshaw elements: localisation and acetylcholinesterase content. *J. Histochem. Cytochem.* **16**, 128–35.

FALCK, B., HILLARP, N.-Å., THIEME, G. & TORP, A. (1962). Fluorescence of catecholamines and related compounds condensed with formaldehyde. *J. Histochem. Cytochem.* **10**, 348–54.

FEDER, N. & WOLF, M. K. (1965). Studies on nucleic acid metachromasy. II. Metachromatic and orthochromatic staining by toluidine blue of nucleic acids in tissue sections. *J. Cell Biol.* **27**, 327–36.

FLICKINGER, C. J. (1966). Methods for handling small numbers of cells for electron microscopy. In *Methods in Cell Physiology*, vol. 2 (ed. D. M. Prescott), pp. 311–21. New York and London: Academic Press.

FORD, E. H. R. & WOOLLAM, D. H. M. (1963). A colchicine, hypotonic citrate, air-drying sequence for foetal mammalian chromosomes. *Stain Technol.* **38**, 271–4.

FORSTER, R. P. & HONG, S. K. (1958). *In vitro* transport of dyes by isolated renal tubules of the flounder as disclosed by direct visualisation. *J. cell. comp. Physiol.* **51**, 259–72.

FRASER ROWELL, C. H. (1963). A general method for silvering invertebrate central nervous systems. *Q. Jl microsc. Sci.* **104**, 81–7.

FREED, J. J. (1969). Microspectrophotometry in the ultraviolet spectrum. In *Physical Techniques in Biological Research* (2nd edn), vol. 3C (ed. A. W. Pollister), pp. 95–167. New York and London: Academic Press.

GALL, J. G. (1966*a*). Chromosome fibers studied by a spreading technique. *Chromosoma* **20**, 221–33.

GALL, J. G. (1966*b*). Techniques for the study of lampbrush chromosomes. In *Methods in Cell Physiology*, vol. 2 (ed. D. M. Prescott), pp. 37–60. New York and London: Academic Press.

GATENBY, J. B. & BEAMS, H. W. (1950). *The Microtomist's Vade-Mecum* (*Bolles-Lee*), 11th edn. London: Churchill.

GAUTHERET, R. J. (1959). *La Culture des Tissus Végétaux*. Paris: Masson.

GEFFEN, L. B., LIVETT, B. G. & RUSH, R. A. (1969). Immunohistochemical localisation of protein components of catecholamine storage vesicles. *J. Physiol., Lond.* **204**, 593–605.

GLAUERT, A. M. (1965a). The fixation and embedding of biological specimens. In *Techniques for Electron Microscopy* (ed. D. H. Kay), 2nd edn, pp. 166–212. Oxford: Blackwell.

GLAUERT, A. M. (1965b). Section staining, cytology, autoradiography and immunochemistry for biological specimens. In *Techniques for Electron Microscopy* (ed. D. H. Kay), 2nd edn, pp. 254–310. Oxford: Blackwell.

GLAUERT, A. M. & PHILLIPS, R. (1965). The preparation of thin sections. In *Techniques for Electron Microscopy* (ed. D. H. Kay), 2nd edn, pp. 213–53. Oxford: Blackwell.

GLEES, P. (1946). Terminal degeneration within the central nervous system. *J. Neuropath. exp. Neurol.* **5**, 54–9.

GOLDACRE, R. J. (1954). A simplified micromanipulator. *Nature, Lond.* **173**, 45.

GOMORI, G. (1955). Preparation of buffers for use in enzyme studies. In *Methods in Enzymology*, vol. 1 (ed. S. P. Colowick and N. O. Kaplan), pp. 138–46. New York: Academic Press.

GOODRICH, E. S. (1942). A new method of dissociating cells. *Q. Jl microsc. Sci.* **83**, 245–58.

GRAY, P. (1954). *The Microtomist's Formulary and Guide*. London: Constable.

GUILLERY, R. W. (1965). Some electron microscopical observations of degenerative changes in central nervous synapses. In *Progress in Brain Research*, vol. 14. *Degeneration patterns in the nervous system* (ed. M. Singer & J. P. Schadé), pp. 57–76. Amsterdam, London and New York: Elsevier.

HAGIWARA, S. & BULLOCK, T. H. (1957). Intracellular potentials in pacemaker and integrative neurons of the lobster cardiac ganglion. *J. cell. comp. Physiol.* **50**, 25–48.

HAINE, M. E. (1961). *The Electron Microscope*. London: Spon.

HALE, A. J. (1958). *The Interference Microscope in Biological Research*. Edinburgh and London: Livingstone.

HALL, C. E. (1966). *Introduction to Electron Microscopy*, 2nd edn. New York: McGraw-Hill.

HALLY, A. D. (1964). A counting method for measuring the volumes of tissue components in microscopical sections. *Q. Jl microsc. Sci.* **105**, 503–17.

HARVEY, R. J. (1968). Measurement of cell volumes by electric sensing zone instruments. In *Methods in Cell Physiology*, vol. 3 (ed. D. M. Prescott), pp. 1–23. New York and London; Academic Press.

HERTWIG, O. & HERTWIG, R. (1879). *Studien zur Blättertheorie I. Die Actinien*, p. 6. Jena: Fischer.

BIBLIOGRAPHY

HEUSER, M. & RAZAVI, L. (1970). A standardised method of peripheral blood culture for cytogenetical studies and its modification by cold temperature treatment. In *Methods in Cell Physiology*, vol. 4 (ed. D. M. Prescott), pp. 477–95. New York and London: Academic Press.

HIRSCH, TH. VON & BOELLARD, J. W. (1958). Methacrylsäureester als Einbettungsmittel in der Histologie. *Z. wiss. Mikrosk.* **64**, 24–9.

HOLMES, W. (1947). The peripheral nerve biopsy. In *Recent Advances in Clinical Pathology* (ed. S. C. Dyke), pp. 402–17. London: Churchill.

HOLTFRETER, J. (1943). Properties and functions of the surface coat in amphibian embryos. *J. exp. Zool.* **93**, 251–323.

HOPWOOD, D., ALLEN, C. R. & McCABE, M. (1970). The reactions between glutaraldehyde and various proteins. An investigation of their kinetics. *Histochem. J.* **2**, 137–50.

HORNE, R. W. (1965). Negative staining methods. In *Techniques for Electron Microscopy* (ed. D. H. Kay), 2nd edn, pp. 328–55. Oxford: Blackwell.

HORRIDGE, G. A. & TAMM, S. L. (1969). Critical point drying for scanning electron microscopic study of ciliary motion. *Science, N.Y.* **163**, 817–8.

INOUÉ, S. (1964). Organisation and function of the mitotic spindle. In *Primitive Motile Systems in Cell Biology* (ed. R. D. Allen & N. Kamiya), pp. 549–98. New York and London: Academic Press.

JAMES, T. H. (ed.) (1966). *The Theory of the Photographic Process*, 3rd edn. New York: Macmillan.

JOHANSEN, D. A. (1940). *Plant Microtechnique*. New York and London: McGraw-Hill.

JONES, B. M. & CUNNINGHAM, I. (1961). Growth by cell division in insect tissue culture. *Expl. Cell Res.* **23**, 386–401.

JUNIPER, B. E., COX, G. C., GILCHRIST, A. J. & WILLIAMS, P. R. (1970). *Techniques for Plant Electron Microscopy*. Oxford and Edinburgh: Blackwell.

KASTEN, F. H. (1967). Cytochemical studies with acridine orange and the influence of dye contaminants in the staining of nucleic acids. *Int. Rev. Cytol.* **21**, 142–202.

KASTEN, F. H. & BURTON, V. (1959). A modified Schiff's solution. *Stain Technol.* **34**, 289.

KAWARAI, Y. & NAKANE, P. (1970). Localisation of tissue antigens on the ultrathin sections with peroxidase-labelled antibody method. *J. Histochem. Cytochem.* **18**, 161–6.

KAY, D. H. (ed.) (1965). *Techniques for Electron Microscopy*, 2nd edn. Oxford: Blackwell.

KERKUT, G. A. & LAVERACK, M. S. (1957). The respiration of *Helix pomatia*, a balance sheet. *J. exp. Biol.* **34**, 97–105.

KIRBY, H. (1950). *Materials and Methods in the Study of Protozoa*. Berkeley and Los Angeles: University of California Press.

KOLB, H. (1970). Organisation of the outer plexiform layer of the primate retina: electron microscopy of Golgi-impregnated cells. *Phil. Trans. R. Soc.* B **258**, 261–83.

KUBITSCHEK, H. E. (1969). Counting and sizing micro-organisms with the Coulter counter. In *Methods in Microbiology*, vol. 1 (ed. J. R.

Norris & D. W. Ribbons), pp. 593–610. London and New York: Academic Press.

LASFARGUES, E. Y. (1957). Cultivation and behaviour *in vitro* of the normal mammary epithelium of the adult mouse. II. Observations on the secretory activity. *Expl. Cell Res.* **13**, 553–62.

LEWIS, P. R. & SHUTE, C. C. D. (1966). The distribution of cholinesterase in cholinergic neurones demonstrated with the electron microscope. *J. Cell Sci.* **1**, 381–90.

LILLIE, R. D. (1965). *Histopathologic Technic and Practical Histochemistry*, 3rd edn. New York, Toronto, Sydney and London: McGraw-Hill.

LOCKWOOD, A. P. M. (1961). 'Ringer' solutions and some notes on the physiological basis of their ionic composition. *Comp. Biochem. Physiol.* **2**, 241–89.

LOUD, A. V. (1962). A method for the quantitative estimation of cytoplasmic structures. *J. Cell Biol.* **15**, 481–7.

LOVELESS, A. (1970). Ambiguous mustard. *Nature, Lond.* **227**, 101.

LUFT, J. H. (1961). Improvements in epoxy resin embedding methods. *J. biophys. biochem. Cytol.* **9**, 409–14.

MCCONNELL, C. H. (1932). The development of the ectodermal nerve net in the buds of *Hydra*. *Q. Jl microsc. Sci.* **75**, 495–510.

MACKINNON, D. L. & HAWES, R. S. J. (1961). *An Introduction to the Study of Protozoa*. Oxford: Clarendon Press.

MAIO, J. J. & SCHILDKRAUT, C. L. (1966). A method for the isolation of mammalian metaphase chromosomes. In *Methods in Cell Physiology*, vol. 2 (ed. D. M. Prescott), pp. 113–30. New York and London: Academic Press.

MARSDEN, C. A. & KERKUT, G. A. (1969). Fluorescent microscopy of 5HT- and catecholamine-containing cells in the central nervous system of the leech *Hirudo medicinalis*. *Comp. Biochem. Physiol.* **31**, 851–62.

MARTIN, L. C. (1966). *The Theory of the Microscope*. London and Glasgow: Blackie.

MASER, M. D., POWELL, T. E. & PHILPOTT, C. W. (1967). Relationships among pH, osmolality, and concentration of fixative solutions. *Stain Technol.* **42**, 175–82.

MEYNELL, G. G. & MEYNELL, E. (1970). *Theory and Practice in Experimental Bacteriology*, 2nd edn. London: Cambridge University Press.

MICHEL, K. (1964). *Die Grundzüge der Theorie des Mikroskops*. Stuttgart: Wissenschaftliche Verlagsgesellschaft.

MILLER, O. L., STONE, G. E. & PRESCOTT, D. M. (1964). Autoradiography of water-soluble materials. In *Methods in Cell Physiology*, vol. 1 (ed. D. M. Prescott), pp. 371–9. New York and London: Academic Press.

MILLONIG, G. & MARINOZZI, V. (1968). Fixation and embedding in electron microscopy. In *Advances in Optical and Electron Microscopy*, vol. 2 (ed. R. Barer & V. E. Cosslett), pp. 251–341. London and New York: Academic Press.

MOOR, H. (1964). Die Gefrier-Fixation lebender Zellen und ihre Anwendung in der Elektronenmikroskopie. *Z. Zellforsch. mikrosk. Anat.* **62**, 546–80.

BIBLIOGRAPHY

MOOR, H. (1971). Recent progress in the freeze-etching technique. *Phil. Trans. R. Soc.* B **261**, 121–31.

MOORE, J. L., WILLIAMS, A. R. & SANDERS, M. (1971). Ultrasonic dispersal of mammalian cell aggregates. *Expl. Cell Res.* **65**, 228–32.

MORENO, G., LUTZ, M. & BESSIS, M. (1969). Partial cell irradiation by ultraviolet and visible light: conventional and laser sources. *Int. Rev. exp. Path.* **7**, 99–137.

MOSCONA, A., TROWELL, O. A. & WILLMER, E. N. (1965). Methods. In *Cells and Tissues in Culture*, vol. 1 (ed. E. N. Willmer), pp. 19–98. London and New York: Academic Press.

MUELLER, G. A., GAULDEN, M. E. & DRANE, W. (1971). The effects of varying concentrations of colchicine on the progression of grasshopper neuroblasts into metaphase. *J. Cell Biol.* **48**, 253–65.

NAIRN, R. C. (ed.) (1969). *Fluorescent Protein Tracing*, 3rd edn. Edinburgh: Livingstone.

NAUTA, W. J. H. (1957). Silver impregnation of degenerating axons. In *New Research Techniques of Neuroanatomy* (ed. W. F. Windle), pp. 17–26. Springfield, Illinois: Thomas.

NAUTA, W. J. H. & EBBESSON, S. O. E. (eds.) (1970). *Contemporary Research Methods in Neuroanatomy*. Berlin, Heidelberg and New York: Springer.

NICOL, J. A. C. (1960). *The Biology of Marine Animals*. London: Pitman.

NIXON, W. C. (1971). The general principles of scanning electron microscopy. *Phil. Trans. R. Soc.* B **261**, 45–50.

NONIDEZ, J. F. (1939). Studies on the innervation of the heart. I. *Am. J. Anat.* **65**, 361–414.

NORRIS, J. R. & RIBBONS, D. W. (eds) (1969, 1970). *Methods in Microbiology*, vols. 1–3. London and New York: Academic Press.

PADAWER, J. (1968). The Nomarski interference-contrast microscope. *Jl R. microsc. Soc.* **88**, 305–49.

PAL, J. (1886). Ein Beitrag zur Nervenfärbetechnik. *Wien med. Jahrb.* **1**, 619–31. (Abstracted in *Z. wiss. Mikrosk.* **4**, 92–6 (1887).)

PANTIN, C. F. A. (1948). *Notes on Microscopical Technique for Zoologists*. London: Cambridge University Press.

PÁRDUCZ, B. (1967). Ciliary movement and coordination in ciliates. *Int. Rev. Cytol.* **21**, 91–128.

PAUL, J. (1970). *Cell and Tissue Culture*, 4th edn. Edinburgh and London: Livingstone.

PEARSE, A. G. E. (1960). *Histochemistry Theoretical and Applied*, 2nd edn. London: Churchill.

PEARSE, A. G. E. (1968). *Histochemistry Theoretical and Applied*, 3rd edn., vol. 1. London: Churchill.

PEASE, D. C. (1964). *Histological Techniques for Electron Microscopy*, 2nd edn. New York and London: Academic Press.

PERRY, R. P. (1964). Quantitative autoradiography. In *Methods in Cell Physiology*, vol. 1 (ed. D. M. Prescott), pp. 305–26. New York and London: Academic Press.

POIRIER, L. J., AYOTTE, R. A. & GAUTIER, C. (1954). Modification of the Marchi technic. *Stain Technol.* **29**, 71–5.

POLANI, P. E. & MUTTON, D. E. (1971). Y fluorescence of interphase nuclei, especially circulating lymphocytes. *Br. med. J.* 1, 138–42.

POLLISTER, A. W., SWIFT, H. & RASCH, E. (1969). Microphotometry with visible light. In *Physical Techniques in Biological Research*, 2nd edn, vol. 3C (ed. A. W. Pollister), pp. 201–51. New York and London: Academic Press.

PRESCOTT, D. M. (1964). Autoradiography with liquid emulsion. In *Methods in Cell Physiology*, vol. 1 (ed. D. M. Prescott), pp. 365–70. New York and London: Academic Press.

PRICE, G. R. & SCHWARTZ, S. (1956). Fluorescence microscopy. In *Physical Techniques in Biological Research*, vol. 3 (ed. G. Oster & A. W. Pollister), pp. 91–148. New York: Academic Press.

PURVIS, M. J., COLLIER, D. C. & WALLS, D. (1966). *Laboratory Techniques in Botany*, 2nd edn. London: Butterworths.

RADIOCHEMICAL MANUAL, THE (1966). Amersham: The Radiochemical Centre.

REEVE, R. M. (1959). A specific hydroxylamine–ferric chloride reaction for histochemical localization of pectin. *Stain Technol.* 34, 209–11.

REISINGER, E. (1960). Vitale Nervenfärbungen bei Plathelminthen und ihre Abhängigkeit vom physiologischen Zustand des Organismus. *Z. wiss. Zool.* 164, 271–93.

REYNOLDS, E. S. (1963). The use of lead citrate at high pH as an electron-opaque stain in electron microscopy. *J. Cell Biol.* 17, 208–12.

RICHARDS, A. G. (1951). *The Integument of Arthropods*. Minneapolis: University of Minnesota Press.

RIEMERSMA, J. C. (1970). Chemical effects of fixation on biological specimens. In *Some Biological Techniques in Electron Microscopy* (ed. D. F. Parsons), pp. 69–99. New York and London: Academic Press.

ROBINOW, C. F. (1970). Staining the *S. pombe* nucleus. In *Methods in Cell Physiology*, vol. 4 (ed. D. M. Prescott), pp. 167–8. New York and London: Academic Press.

ROGERS, A. W. (1967). *Techniques of Autoradiography*. Amsterdam, New York and London: Elsevier.

ROOTS, B. I. (1955). The water relations of earthworms. *J. exp. Biol.* 32, 765–74.

ROSE, G. G. (1967). The circumfusion system for multipurpose culture chambers. 1. Introduction to the mechanics, techniques and basic results of a 12-chamber (*in vitro*) closed circulatory system. *J. Cell Biol.* 32, 89–112.

ROSS, K. F. A. (1961). The immersion refractometry of living cells by phase contrast and interference microscopy. In *General Cytochemical Methods*, vol. 2 (ed. J. F. Danielli), pp. 1–60. New York: Academic Press.

ROSS, K. F. A. (1967). *Phase Contrast and Interference Microscopy for Cell Biologists*. London: Edward Arnold.

ROTH, L. J. & STUMPF, W. E. (eds.) (1969). *Autoradiography of Diffusible Substances*. New York and London: Academic Press.

RUDE, S., COGGESHALL, R. E. & VAN ORDEN, L. S. (1969). Chemical and ultrastructural identification of 5-hydroxytryptamine in an identified neuron. *J. Cell Biol.* 41, 832–54.

BIBLIOGRAPHY

RUTHMANN, A. (1970). *Methods in Cell Research*. London: Bell.

SALPETER, M. M. (1966). General area of autoradiography at the electron microscope level. In *Methods in Cell Physiology*, vol. 2 (ed. D. M. Prescott), pp. 229–53. New York and London: Academic Press.

SASS, J. E. (1958). *Botanical Microtechnique*, 3rd edn. Ames, Iowa: Iowa State College Press.

SCHULTZE, B. (1969). Autoradiography at the cellular level. In *Physical Techniques in Biological Research*, 2nd edn, vol. 3B (ed. A. W. Pollister). New York and London: Academic Press.

SCOTT, J. E. (1967). On the mechanism of the methyl green–pyronin stain for nucleic acids. *Histochemie* 9, 30–47.

SHARMA, A. K. & SHARMA, A. (1965). *Chromosome Techniques – Theory and Practice*. London: Butterworths.

SHEA, J. R. (1970). A method for *in situ* cytophotometric estimation of absolute amount of ribonucleic acid using azure B. *J. Histochem. Cytochem.* 18, 143–52.

SJÖSTRAND, F. S. (1967). *Electron Microscopy of Cells and Tissues*, vol. 1, *Instrumentation and Techniques*. New York and London: Academic Press.

SLAYTER, E. M. (1970). *Optical Methods in Biology*. New York, London, Sydney and Toronto: Wiley–Interscience.

SMALL, E. B. & MARSZALEK, D. S. (1969). Scanning electron microscopy of fixed, frozen and dried protozoa. *Science, N.Y.* 163, 1064–5.

SMITH, R. E. (1970). Comparative evaluation of two instruments and procedures to cut non-frozen sections. *J. Histochem. Cytochem.* 18, 590–1.

STANILAND, L. N. (1952). *The Principles of Line Illustration*. London: Burke.

STEEDMAN, H. F. (1947). Ester wax: a new embedding medium. *Quart. Jl microsc. Sci.* 88, 123–33.

STEEDMAN, H. F. (1960). *Section Cutting in Microscopy*. Oxford: Blackwell.

STELL, W. K. (1965). Correlation of retinal cytoarchitecture and ultrastructure in Golgi preparations. *Anat. Rec.* 153, 389–98.

STEVENS, A. R. (1966). High resolution autoradiography. In *Methods in Cell Physiology*, vol. 2 (ed. D. M. Prescott), pp. 255–310. New York and London: Academic Press.

STONE, A. L. & BRADLEY, D. F. (1961). Aggregation of acridine orange bound to polyanions: the stacking tendency of deoxyribonucleic acids. *J. Am. chem. Soc.* 83, 3627–34.

STRAUSFELD, N. J. & BLEST, A. D. (1970). Golgi studies on insects. Part I. The optic lobes of Lepidoptera. *Phil. Trans. R. Soc. B* 258, 81–134.

STRETTON, A. O. W. & KRAVITZ, E. A. (1968). Neuronal geometry: determination with a technique of intracellular dye injection. *Science, N.Y.* 162, 132–4.

SUMNER, A. T., ROBINSON, J. A. & EVANS, H. J. (1971). Distinguishing between X, Y and YY-bearing human spermatozoa by fluorescence and DNA content. *Nature, New Biol.* 229, 231–3.

SWANK, R. L. & DAVENPORT, H. A. (1935). Chlorate–osmic–formalin method for staining degenerating myelin. *Stain Technol.* 10, 87–90.

TYRODE, M. V. (1910). The mode of action of some purgative salts. *Archs. int. Pharmacodyn. Thér.* **20**, 205–23.

WANSON, J. C. (1964). Techniques de coupes semi-fines. Adaptation pour le microscope à lumière visible. *J. Microscopie* **3**, 413–26.

WEIBEL, E. R. (1969). Stereological principles for morphometry in electron microscopic cytology. *Int. Rev. Cytol.* **26**, 235–302.

WEIBEL, E. R. & ELIAS, H. (1967). *Quantitative Methods in Morphology.* Berlin, Heidelberg and New York: Springer.

WEIS-FOGH, T. (1956). Tetanic force and shortening in locust flight muscle. *J. exp. Biol.* **33**, 668–84.

WEST, S. S. (1969). Fluorescence microspectrophotometry of supravitally stained cells. In *Physical Techniques in Biological Research*, 2nd edn, vol. 3C (ed. A. W. Pollister), pp. 253–321. New York and London: Academic Press.

WIGGLESWORTH, V. B. (1950). A new method for injecting the tracheae and tracheoles of insects. *Q. Jl microsc. Sci.* **91**, 217–24.

WIGGLESWORTH, V. B. (1959). A simple method for cutting sections in the 0.5 to 1 μ range, and for sections of chitin. *Q. Jl microsc. Sci.* **100**, 315–20.

WIGGLESWORTH, V. B. (1971). Bound lipid in the tissues of mammal and insect: a new histochemical method. *J. Cell Sci.* **8**, 709–25.

WOOD, J. G. & BARRNETT, R. J. (1964). Histochemical demonstration of norepinephrine at a fine structural level. *J. Histochem. Cytochem.* **12**, 197–209.

WOODS, R. I. (1969). Acrylic aldehyde in sodium dichromate as a fixative for identifying catecholamine storage sites with the electron microscope. *J. Physiol. Lond.* **203**, 35 P–36 P.

YAMASAKI, T. & NARAHASHI, T. (1959). The effects of potassium and sodium ions on the resting and action potentials of the cockroach giant axon. *J. Insect Physiol.* **3**, 146–58.

YOUNG, J. Z. (1933). The preparation of isotonic solutions for use in experiments with fish. *Pubbl. Staz. zool. Napoli* **12**, 425–31.

YUNIS, J. J. (ed.) (1965). *Human Chromosome Methodology.* New York and London: Academic Press.

ZWILLING, E. (1954). Dissociation of chick embryo cells by means of a chelating compound. *Science, N.Y.* **120**, 219.

Index

INDEX

interference microscope, 4
invertebrates: fixatives for, 22, 23; narcotization of, 16, 17; nervous systems of, 93; stain for, 46
iodine in potassium iodide solution, 55, 84, 90
iris diaphragm, 2, 3
iron haematoxylin stain (Heidenhain), 39, 41, 75, 86, 91
iron trioxyhaematin stain (Hansen), 40, 42
isotopes, radioactive, 68–70

Janus green stain, 14, 84

Knop solution, for plants, 105

lactophenol/cotton blue stain, 40, 47
Lambert–Beer law of light absorption, 8, 9
laser light, 15
lead salts, as stains in electron microscopy, 59, 64
Leishman stain, 101
lenses, cleaning of, 3
leucocytes, 102–3
light green stain, 39
lignin, stains for, 46, 55, 56
lipids: in fixation, 20, 22; in freeze substitution, 27; histochemical work on, 35; stains for, 53–4; unmasking of 'bound', 54
living material, 3, 4, 13–14; manipulation of, 14–15; mitotic spindles in, 79; protozoa, 83–4
locust Ringer solution, 105
Lugol's iodine, 84
lymphocytes, division of, 75

McConnell method for staining nervous systems, 94–5
maceration: of insect material, 90; of plant tissues, 76; of tissues to obtain cells, 19
macrophages, stain for, 14
magnesium chloride solution, as narcotic, 16
magnification, 2
Mallory stain, 39, 43, 86, 93
mammalian tissues: of central nervous system, 93; fixative for electron microscopy of, 60; salines for, 107
marine organisms, fixatives for, 22, 23, 25
mast cells, stain for, 45

Mayer's albumen, adhesive, 31, 32, 86
menthol, as narcotic, 16
mercuric chloride method, for staining nerves, 94
mercury arc lamp, for fluorescence microscopy, 6
metachromasy, 48–9, 55
metaphase, arrest of mitosis at, 75–6
methacrylate embedding media, 25, 34, 35, 70
methyl benzoate, as antemedium, 29
methyl blue/eosin stain (Mann), 39, 45, 91
methyl cellulose, immobilization by, 13, 84
methyl green, 84
methyl green/pyronin stain, 51–2, 73, 75
methylene blue stain, 39, 48, 65, 94
micromanipulators, 15
micrometer, 7
micropipettes, 15
microspectrophotometry, 8–9
microtome, 30–1; freezing, 36–7; sliding (base-sledge), 35; ultra-, 33, 48, 62–3, 65
microtubules, 67
Millon reaction, for tyrosine in proteins, 8, 52
mitochondria, stain for, 14, 84
mounting of sections and smears, 57–8
MS 222, narcotic, 16
mucopolysaccharides, stains for, 55, 65
mucus, stains for, 43, 45
muscle cells: birefringent, 5; stains for, 45, 91
myelin, stains for, 95–6, 97–8

naphthol yellow stain, 8
narcotisation, 13, 16–17, 84
Nauta method for staining degenerating axons, 98
Navashin fluids, fixatives, 24
necrosis in tissues, stain for, 44, 45
negative staining, 66–7
nervous systems, 93–6; studies of degeneration in, 96–8; synapses in, 99
neutral red stain, 14, 39
nigrosin method for examining bacteria, 80
Nissl substance, 97
nitric acid, decalcification by, 37
Noland method for staining cilia, etc., 84